BASIC CONCEPTS IN CHEMISTRY

Reactions and Characterization of Solids

SANDRA E DANN

Loughborough University

WILEY-INTERSCIENCE

RS•C

ROYAL SOCIETY OF CHEMISTRY

Cover images © Murray Robertson/visual elements 1998–99, taken from the
109 Visual Elements Periodic Table

For ordering and customer service, call 1-800-CALL-WILEY.

Library of Congress Cataloging-in-Publication Data:
Library of Congress Cataloging-in-Publication Data is available.
ISBN: 0-471-22481-2

Typeset in Great Britain by Wyvern 21, Bristol
Printed and bound by Polestar Wheatons Ltd, Exeter

10 9 8 7 6 5 4 3 2 1

Preface

Materials is one area of research which brings together all the physical sciences: chemists to synthesize and characterize the solids, physicists to investigate their properties and engineers to fabricate them into useful devices.

Since the last decade, when I studied my degree, the materials subject has become an essential part of a chemistry degree, as the relationship between chemical structure and inherent physical properties has been recognized.

This book is designed as an introductory text to the synthesis and characterization of materials in the solid state, which are rather different to their solution-made, mononuclear counterparts. The text covers the salient information in a way which I would have found useful when I was studying for my degree. In addition, the last chapter introduces some recent discoveries, *e.g.* the superconducting oxides, which have been developed owing to the collaboration of chemists, physicists and engineers.

I have many people to thank in the conception and writing of this text. Firstly, Professor J. Derek Woollins for his invitation and, not least of all, Dr Mike Webster for his endless help and encouragement when I started teaching at Southampton. I would also like to thank Dr Paul Andrews, Deana McDonagh-Philp and Jennifer Armstrong for their proofreading and help, when I was less than gracious at the time, and a generation of Southampton and Loughborough University students, who were unknowing guinea-pigs for the questions in this book. Finally, I would like to thank my parents, for finding the money to send me to university, and the continual support of my family.

S. E. DANN
Loughborough

BASIC CONCEPTS IN CHEMISTRY

EDITOR-IN-CHIEF

Professor E W Abel

EXECUTIVE EDITORS

Professor A G Davies
Professor D Phillips
Professor J D Woollins

EDUCATIONAL CONSULTANT

Mr M Berry

This series of books consists of short, single-topic or modular texts, concentrating on the fundamental areas of chemistry taught in undergraduate science courses. Each book provides a concise account of the basic principles underlying a given subject, embodying an independent-learning philosophy and including worked examples. The one topic, one book approach ensures that the series is adaptable to chemistry courses across a variety of institutions.

TITLES IN THE SERIES

Further information about this series is available at www.wiley.com/go/wiley-rsc

Contents

1
Solid State Structure

Although the majority of this book will be concerned with crystalline inorganic materials, background knowledge of the different types of structures and bonds is essential for an understanding of the properties of *any* material.

Until recently, the structures of inorganic systems have been relatively mysterious, since the large number of possible coordination numbers and geometries, as well as possible combinations of elements, made these materials more difficult to characterize than their organic counterparts. This was particularly true of glassy and disordered materials, which could not be investigated using standard techniques such as X-ray diffraction.

Aims

This chapter deals with the fundamental concepts of crystal classification, symmetry and inorganic crystal chemistry. By the end of this chapter you should be able to

- Understand the differences between crystalline and amorphous materials
- Draw, as projections, the simple structures derived from close-packing and 'atom count' to show the number of formula units in each unit cell
- Use ionic radius ratios to predict likely structures

1.1 Types of Solid

Many industrially and technologically important materials are solids. In addition to the magnetic, optical or mechanical properties and anything else which makes them interesting, a chemist would also want to know about their **structure**: which atoms and/or ions are involved, where they are relative to each other and how they are bonded together.

All materials fall into two groups:

- Crystalline: long-range order ($>10^3$ molecules)
- Non-crystalline   small particles with no long-range order
  (amorphous):   (100 Å, where 1 Å = 10^{-8} cm)

Nearly all materials can be prepared in the amorphous state. Interest in amorphous materials has grown with the development of techniques which can be used to characterize them. Glasses are a special type of amorphous material which melts at the **glass transition temperature**; here the rate of cooling and the viscosity of the liquid, when the material solidifies, are too great to allow the atoms to rearrange into an ordered crystalline state.

Although glasses and other amorphous materials are difficult to characterize, this does not make them useless. For example, amorphous silicon can be used to transform solar energy into electricity, and glasses obviously find applications in window panes, bottles, drinking glasses, *etc.*

In the crystalline state, regular atomic order persists over distances which are very large in comparison with interatomic distances. However, even in the most perfect of crystals there are some small and usually random departures from regularity. These imperfections result in minor changes to physical properties, such as resistance and conductivity, but are a feature of solid state materials in general. It is noteworthy that reactions of materials often involve the whole crystal lattice, imperfections and all, rather than just the atoms of the 'pure' substance.

Crystals can be formed in numerous ways, including cooling from molten salts and deposition from vapours. Whether the material is obtained as a single crystal or a polycrystalline mass depends on the conditions used. The polycrystalline mass often has random directional properties, so it is better to characterize physical properties using a single crystal rather than the mass. However, the external symmetry of even the smallest crystallite is characteristic of that particular substance, and is normally related to the arrangement of the atoms in the crystal. The macroscopic appearance of the crystal therefore gives information on the microscopic arrangement of its atoms.

1.2   Properties of Large Crystals

The word crystal is derived from the ancient Greek word for ice. The ancient Greeks believed that quartz (SiO_2) was water so deeply frozen it would never melt. Although quartz was found much later to contain silicon and oxygen rather than hydrogen and oxygen, the name 'crystal' is still used as a general term today.

1.2.1 Characteristics of Macroscopic Crystals

Some properties of large crystals found in museums or grown in the laboratory are given below:

- Flat faces (to a good approximation)
- Relative *sizes* of faces can vary from crystal to crystal of the same substance
- The *angle* between similar faces is found experimentally to be a constant
- Crystals break (cleave) parallel to certain well-defined faces

Although the angles between faces remain constant, the relative sizes of faces may vary from crystal to crystal. This behaviour is described as **crystal habit**, and the growth of the same substance from different solutions can result in different shapes. For example, sodium chloride crystals grown in water are cubic, but if urea is added to the water they become octahedral (a regular octahedron and a cube have the same overall symmetry).

Polymorphism occurs when different structures can occur for the same chemical formula. The atoms are the same but their atomic arrangement differs between the structures. Although ultimately thermodynamics (the minimum Gibbs free energy) dictates which of the probable structures is formed at a particular temperature and pressure, other factors such as electrostatic interaction mean that a variety of different structures is possible. Polymorphism applies not only to elements, *e.g.* black and red phosphorus, but also to compounds such as calcium carbonate, which can exist in a number of forms including calcite, aragonite and vaterite.

1.3 Crystal Lattices

A classification of crystals which is based on bonding is useful in understanding structure–property relations in solids. Five types of solid are readily defined by considering the bonding in them: ionic, covalent, metallic, molecularly bonded and hydrogen bonded. The last two are concerned with molecules rather than with lattices and will not be discussed here. They are discussed in detail in other texts (*e.g.* Rao and Gopalakrishnam[1]). Important properties of the particular lattices are summarized in Table 1.1.

1.3.1 Ionic Crystals

Ionic crystals are formed between highly electropositive and highly electronegative elements when electron transfer has occurred between the atoms, resulting in oppositely charged ions with closed shell (octet) electronic configurations. Ionic crystals such as potassium chloride

Table 1.1 Important properties of particular lattices

Type	Particles	Characteristics	Examples
Ionic	Ions	Brittle, insulating, high melting point	NaCl, CaF_2
Covalent	Atoms	Rigid, high melting point, non-conducting (pure)	Diamond SiC
Metallic	Positive ions in a 'sea' of electrons	High conductivity	Na, Au

consist of spherical ions, where charged ions are surrounded by ions of the opposite charge. This results in highly symmetrical structures in which like charges are as far apart as possible.

1.3.2 Covalent Crystals

When the elements in a crystalline material have similar electronegativities, bond formation occurs through the sharing of electrons, each atom contributing one electron to the bond. The force of attraction between two atoms of similar electronegativity arises from the overlap of atomic orbitals and the consequent net lowering of energy. This results in the formation of an electron pair, where the electrons are shared between neighbouring atoms. Typical examples of covalently bonded solids include those in Group 14, where sharing of the four s^2p^2 electrons achieves a closed shell electronic configuration, *e.g.* SiC. Hybridization of the atomic orbitals results in a tetrahedral arrangement of the four covalent bonds.

1.3.3 Metallic Crystals

Metals are significantly different from non-metallic solids both in structure and physical properties. They usually crystallize with dense structures (see Section 1.5) with large coordination numbers (eight to twelve). Typically they consist of positively charged ions in a 'sea' of electrons, in which the outer electrons of the metal atoms are free to move, hence giving high electrical and thermal conductivity. The properties of metallic crystals cannot be accounted for in terms of localized bonds or electron clouds, and are best described by a 'delocalized electron' approach.

A simple explanation for the many characteristic features of the metallic state is given by 'free-electron' theory. In metallic crystals the atoms are assumed to take part collectively in bonding, where each atom provides electrons from outer electron energy levels to the bond. The crystal is considered to be 'held together' by electrostatic interaction between the resulting cations and the free electrons. Within the boundaries of the crystal, the free electrons move under a constant potential, while at the boundaries there is a large potential difference which prevents escape of the electrons.

Polymorphism in metals is common, often occurring as the temperature of the metal changes, and the different forms are normally (though not exclusively) labelled α, β, γ ... with increasing temperature. In some cases, metals can revert to a lower temperature form at high temperature. For example α-Fe is stable up to 906 °C and then alters to γ-Fe at 1401 °C before reverting to α-Fe at 1530 °C. In this case, β-Fe is not stable under normal conditions and is only found at high pressure.

1.4 Unit Cells and Symmetry Elements

The properties observed for large macroscopic crystals suggested to early observers that crystals were made up by the regular repetition in space of the *same* unit of structure. Each **unit cell** displays the full symmetry of the structure.

A three-dimensional unit cell is described as a small volume defined by six faces. Each face is a parallelogram and there are three identical pairs. The bulk material is made up by translation (displacement) along the unit cell axes by an integral number of lengths of the axes. For example, Figure 1.1 gives a picture of a two-dimensional repeating pattern. For a two-dimensional pattern the unit cell is a parallelogram with two pairs

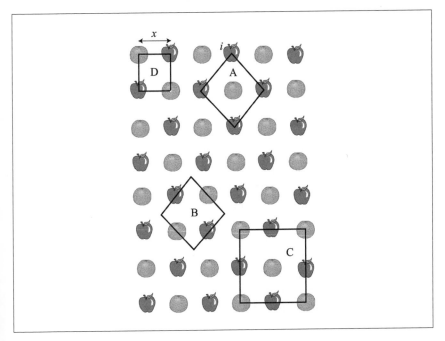

Figure 1.1 Two-dimensional repeating lattice

of identical sides which must show the full symmetry of the structure. Although any of the cells A, B or C could be chosen, the unit cell D is not a suitable choice as translation by the distance x translates an apple on to an orange. Movement of the other cells by a distance equal to their cell edges always maps apple on to apple or orange on to orange. The fact that the points chosen here are apples and oranges is irrelevant, as unit cells and symmetry apply to *any set of points* and do not necessarily relate directly to atoms. Any identical array of repeating points can be referred to as a lattice and hence can show symmetry and possess a unit cell.

Although cells A, B and C all show the full symmetry of this lattice, C is twice the size of the other cells. This introduces the principle of *point counting* and sharing to calculate how many lattice points a cell contains. In order to count points, the number of cells each point is shared between must be considered. Point i in cell A is on a corner, and if cells were drawn in both directions it could be seen that i is shared between four cells. This means that i is only worth one quarter of a whole point to the cell A. If we repeat this procedure for the whole of A, this generates one whole orange and one whole apple inside A. Thus for apples: $4 \times \frac{1}{4} = 1$ apple; for oranges $= 1 \times 1 = 1$ orange

Worked Problem

Q Repeat this procedure for cell C.

A Repeating this procedure for C produces two apples and two oranges in a unit cell. A unit cell which contains only one lattice point of a particular type is known as **primitive**. So here A is a primitive unit cell and C is non-primitive as it contains two lattice points (two apples and two oranges).

For a three-dimensional unit cell, the lengths (**cell parameters**) and angles are conventionally given the symbols a, b, c, α, β and γ, and are defined as shown in Figure 1.2. In three-dimensional unit cells, atom sharing leads to the proportional quantities given in Table 1.2.

The unit cell shown in Figure 1.2 has no symmetry in that the cell parameters and angles may take any values. All the cell parameters (a, b, c) have different lengths and all the angles are not right angles (90°). Increasing the level of symmetry produces relationships between the various cell parameters. For example, if a, b and c are all the same length and all the angles are right angles, then the shape of the unit cell is cubic. Seven crystal classes are obtained by the different possible combinations of cell parameters and angles which are given in Table 1.3 and illustrated in Figure 1.3.

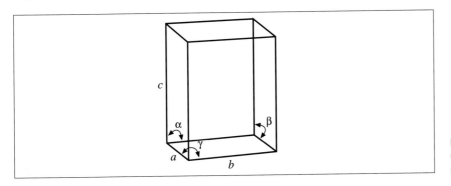

Figure 1.2 Three-dimensional unit cell with conventional lattice parameters

Table 1.2 Proportional quantities in three-dimensional unit cells

Position point	Fraction of lattice
Corner	1/8
Edge	1/4
Face	1/2
Inside Cell	1

Worked Problem

Q Take eight sugar cubes, place the first four in a two-by-two formation. Place the remaining four on top of the first four to form a larger cube. Imagine a sphere at the centre of the large cube, on a corner of each of the sugar cubes. One eighth of the sphere lies within each sugar cube.

(i)   Repeat the process for a sphere which lies on the face of a sugar cube.

(ii) Locate the sphere on the edge of a sugar cube.

(iii) Place the sphere inside a sugar cube.

How much of the sphere lies within each sugar cube in the three tasks?

A (i) One half.

  (ii) One quarter.

  (iii) One.

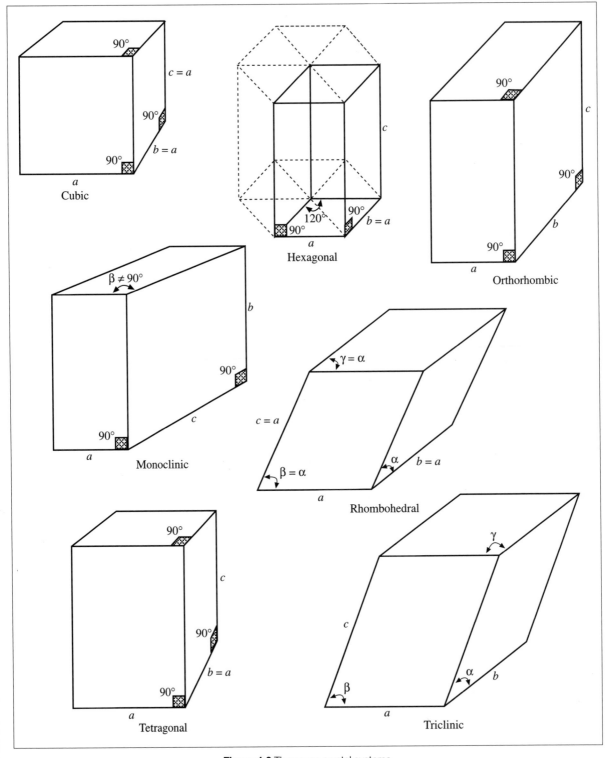

Figure 1.3 The seven crystal systems

Table 1.3 Crystal classes

Cell parameter relationship		Crystal class
$a = b = c$	$\alpha = \beta = \gamma = 90°$	Cubic
$a = b \neq c$	$\alpha = \beta = \gamma = 90°$	Tetragonal
$a \neq b \neq c$	$\alpha = \beta = \gamma = 90°$	Orthorhombic
$a \neq b \neq c$	$\alpha = \gamma = 90°, \beta \neq 90°$	Monoclinic
$a \neq b \neq c$	$\alpha \neq \beta \neq \gamma \neq 90°$	Triclinic
$a = b \neq c$	$\alpha = \beta = 90°, \gamma = 120°$	Hexagonal
$a = b = c$	$\alpha = \beta = \gamma \neq 90°$	Trigonal (rhombohedral)

Three-dimensional figures such as these can be difficult to draw in an unambiguous way, so these diagrams are simplified by drawing a projection. In this, the structure is drawn by viewing it from a particular direction, normally along one of the unit cell axes. Viewing along an axis means that one of the faces of the unit cell is nearest to the viewer, and the distance of each atom from that face is written on the projection; the distance is measured as a fraction of the cell parameter along the axis of viewing. For example, Figure 1.4 shows a cube containing a lattice point at each corner and one at the centre of the cell and its two-dimensional projection. If we view the cube from the top face and consider first the atoms on the corners, for each corner there is one on the top face at 0 and one at the bottom at 1 (*i.e.* the whole unit cell away). The atom at the centre of the cube is half-way down the unit cell, so it is shown at a distance of $\frac{1}{2}$ on the projection.

This procedure is rather limited as it can only really be useful for unit cells with at least one 90° angle. A more universal approach involves the use of **fractional coordinates**. With these, each atom in the unit cell is given an x, y and z coordinate, where the atom is positioned at $(x/a, y/b, z/c)$ with respect to the origin of the unit cell. For example, in Figure 1.4 the atom at the centre of the unit cell would have a fractional coordinate of $(\frac{1}{2}, \frac{1}{2}, \frac{1}{2})$ corresponding to a translation of half way along x, half way along y and half way along z with respect to the unit cell origin.

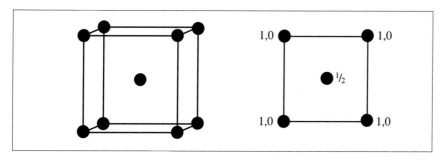

1,0   1,0

$\frac{1}{2}$

1,0   1,0

Figure 1.4 Orthographic projection of a body-centred cubic unit cell

In addition to different types of crystal system there are also different types of lattice within those crystal systems, which correspond to specific arrangements of the atoms/ions within them. As discussed earlier, the two-dimensional system with the simplest sort of lattice which contains only one lattice point, is termed **primitive**. Similarly, for each three-dimensional crystal system there is always a primitive unit cell which consists of atoms located at the corners of the particular parallelepiped (*i.e.* a solid figure with faces which are parallelograms). For example, Figure 1.5 shows primitive lattices (symbol **P**) for both tetragonal and hexagonal systems.

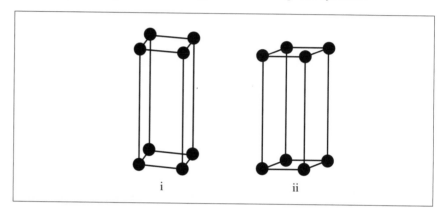

Figure 1.5 (i) Primitive tetragonal ($a = b \neq c$, $\alpha = \beta = \gamma = 90°$) and (ii) primitive hexagonal ($a = b \neq c$, $\alpha = \beta = 90°$, $\gamma = 120°$) unit cells

For the remaining lattice types, as well as the translational symmetry from these corner points there is additional symmetry within the unit cell. The simplest of these is the **body-centred** example seen in Figure 1.4, where there is also an atom at the centre of the unit cell (with fractional coordinates ($1/2$, $1/2$, $1/2$)). For a lattice to be body-centred therefore, if an atom or ion is placed on x, y, z there must be an identical one placed at $x+1/2$, $y+1/2$, $z+1/2$. A body centred lattice is given the symbol **I**.

In additional to body-centred, there are also two possible types of **face-centred** unit cells. A lattice where all the faces have a centrally placed atom is given the symbol **F**. If only one pair of faces is centred, then the lattice is termed **A**, **B** or **C** depending on which face the centring occurs. For example, if the atom or ion lies on the face created by the a and b axes, the lattice is referred to as C-centred. Examples of face-centred lattices are given in Figure 1.6.

Not all types of lattice are allowable within each crystal system, because the symmetrical relationships between cell parameters mean a smaller cell could be drawn in another crystal system. For example a C-centred cubic unit cell can be redrawn as a body-centred tetragonal cell. The fourteen allowable combinations for the lattices are given in Table 1.4. These lattices are called the **Bravais** lattices.

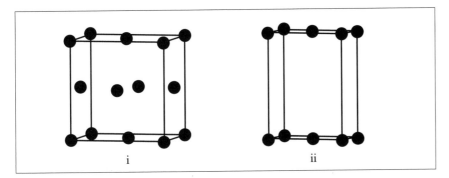

i   ii

Figure 1.6 (i) Face-centred cubic (F) and (ii) face-centred orthorhombic (C)

Table 1.4 Bravais lattices

Crystal system	Lattice types
Cubic	P, I, F
Tetragonal	P, I
Orthorhombic	P, C, I, F
Monoclinic	P, C
Triclinic	P
Hexagonal	P
Trigonal (rhombohedral)	P*

* Often the primitive rhombohedral lattice is referred to as R.

1.5 Structures Formed by the Close Packing of Spheres

In molecular chemistry, octahedral and tetrahedral molecules can link together to form clusters. One way in which we can visualize solid structures is as semi-infinite lattices built up from unlimited numbers of tetrahedral and octahedral units (atoms or ions). We can consider polyhedra packing in various ways by sharing edges or vertices to form dense structures. This procedure is covered in many texts,[2,3] and will not be described in detail here.

Another way in which many solids can be described is by thinking of atoms as hard spheres which can pack together as tightly as geometry will allow. This close-packing procedure creates holes which can be then filled by other atoms, and this principle can be used to explain many simple structures.

If identical hard spheres such as ball bearings are placed in a flat box to form a single layer, each sphere is soon surrounded by six others. This arrangement provides the most efficient packing model. In general, in structural chemistry the number of nearest neighbours around an atom is known as its **coordination number**. For these ball bearings in a single layer the coordination number is six (Figure 1.7).

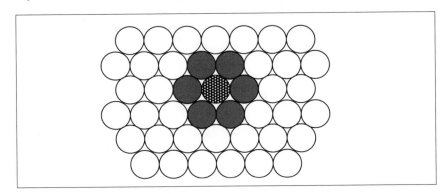

Figure 1.7 Packing of hard spheres in a single layer

However, once we have filled the bottom of the box, the ball bearings start forming the next layer. The most efficient packing arrangement involves the newly added ball bearings falling into the depressions created by the first layer (Figure 1.8).

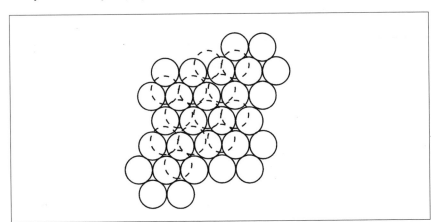

Figure 1.8 Second layer of spheres lying in depressions created by the first

This procedure creates two sorts of hole as the round ball bearings do not fill all the space: a tetrahedral hole which is surrounded by four spheres and an octahedral hole which is surrounded by six spheres (Figure 1.9).

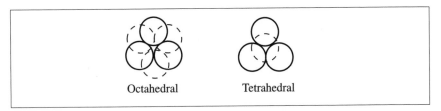

Octahedral   Tetrahedral

Figure 1.9 Tetrahedral hole created by four spheres and octahedral hole created by six

Once the second layer is completed, the third layer can be laid down in two different ways to create two different structural **polytypes**. When the layer is placed with all atoms directly above those of the first layer, the result is an **ABA** structure which is known as **hexagonal close packing** (hcp). Alternatively, if the third layer is displaced so that each atom is above a hole in the first layer, the resulting **ABC** structure is called **cubic close packing** (ccp) (Figure 1.10). The unit cell of the latter is actually a face-centred cube. In both cases the coordination number of any sphere is 12, because the atom is surrounded by six atoms in its own layer with three atoms in the layer above and three in the layer below.

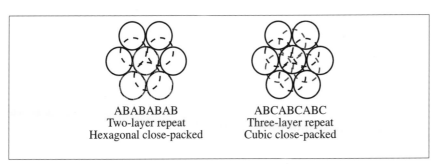

ABABABAB
Two-layer repeat
Hexagonal close-packed

ABCABCABC
Three-layer repeat
Cubic close-packed

Figure 1.10 Hexagonal (two-layer) and cubic (three-layer) close packing

The hexagonal structure shown in Figure 1.11 shows a section of the unit cell. The reason for the *hexagonal* label is clear: hexagons share vertical faces to form the three-dimensional lattice. The unit cell for hexagonal close packing is, however, only a third of this and is much less obvious. The hexagonal unit cell contains eight atoms on the corners (each counting one-eighth by the rules of atom counting) and one atom inside the unit cell. This means the number of atoms in a hexagonal unit cell is two.

The cubic part of the cubic close-packed structure is more difficult to see, but can be visualized by considering the alternating layers of six

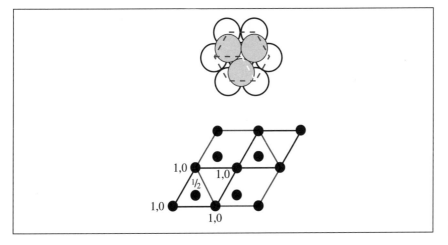

Figure 1.11 Hexagonal close-packed unit cell

atoms which, with two further atoms (one above and one below), produce the cube shown in Figure 1.12. In this case, counting of atoms gives four per unit cell. This diagram shows that the unit cell for cubic close packing is a face-centred cube.

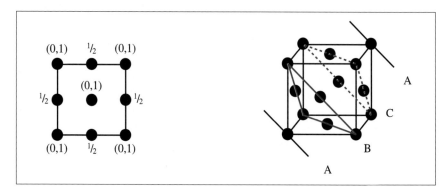

Figure 1.12 Cubic close-packed unit cell

Many ionic crystals can also be described in terms of filling the holes (sometimes termed **interstices**) in the close-packed structure which are created by packing the spheres. In a way similar to atom sharing, the interstices or holes are shared. A projection of the positions of the holes in cubic and hexagonal close packing is shown in Figure 1.13. For each atom or ion, one octahedral hole and two tetrahedral holes are created by the packing of the atoms in both types of close packing.

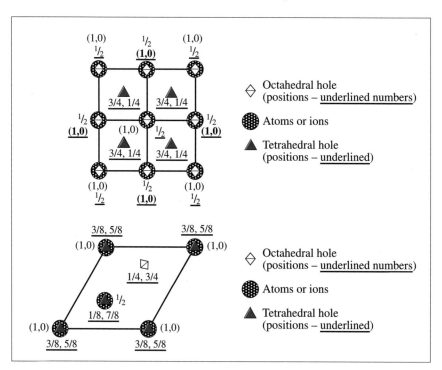

Figure 1.13 Position of octahedral and tetrahedral holes in hcp and ccp

Worked Problem

Q Use the atom/hole sharing principle to calculate the ratio of octahedral hole/tetrahedral hole/atoms in a cubic close-packed structure

A There are 12 octahedral holes on the edge of the cell (each counting a quarter) and one in the centre (counting one). All eight of the
tetrahedral holes lie within the unit cell, counting one each. Finally,
there are eight atoms on the corners (counting one-eighth) and six
atoms on the faces (counting a half). The ratio of octahedral
hole:tetrahedral hole:atoms is then 4:8:4, which gives an empirical
ratio of 1:2:1. For any close-packed structure the ratio would be 1:2:1

The structures of binary ionic compounds (those consisting of two elements) can be explained by considering the larger ions (normally the
anions) forming the lattice and the smaller ions (normally the cations) filling the holes. This leads to a number of structures which are considered
typical of a particular group. For example, the sodium chloride structure
is found in more than 60 binary compounds which share the same packing arrangement.

1.5.1 Sodium Chloride (NaCl) (Cubic Close Packed, Octahedral Holes Fully Occupied)

The 'rock-salt' or 'halite' structure is one of the most simple and well-
known structures, with many halides and oxides showing a similar
arrangement. A three-dimensional picture and projection of the structure
is shown in Figure 1.14. All the octahedral holes created by the ions are
filled, creating a ratio of 4Na:4Cl by atom/hole counting. This is characteristic of all face-centred cubic lattices; four formula units (*e.g.* 4NaCl)
are present in the unit cell.

Another way of visualizing this structure is two interpenetrating face-
centred cubes, where one is displaced by half a unit cell relative to the
other. Each sodium ion is surrounded by six chlorides, and each chloride
is surrounded by six sodium ions. This structure is often referred to as a
6:6 structure from the coordination number of the two ions.

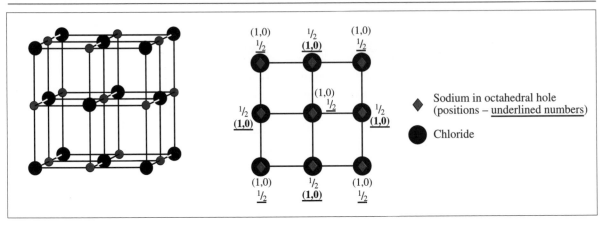

Figure 1.14 Sodium chloride (halite) structure

1.5.2 Calcium Fluoride (Cubic Close Packed, All Tetrahedral Holes Fully Occupied)

This is another very common structure with more than 50 examples, including UO_2, CeO_2 and TiH_2. As there are twice as many fluoride ions in the unit cell as calcium ions, the anions must logically fill the tetrahedral holes and the calcium ions form the lattice to give the correct ratio of ions. Cations and anions are therefore the opposite way round to what would normally be expected in a binary ionic compound (see above). Ca:F has a ratio of 4:8 (or 1:2) since there are twice as many tetrahedral holes as atoms in the structure. As expected, the fluoride ion must have a coordination number of four as it is in a tetrahedral hole. A projection and diagram are given in Figure 1.15.

Worked Problem

Q What is the coordination number of the calcium ions in CaF_2?

A As the lattice is continuously repeating, each calcium must have eight fluorides as nearest neighbours.

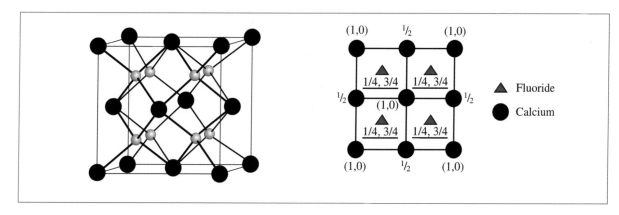

1.5.3 Zinc Blende (Cubic Close Packed, Tetrahedral Holes Half Occupied)

Figure 1.15 Calcium fluoride (fluorite) structure

The structure of zinc blende is given in Figure 1.16. This is a relatively common structure, with more than 30 examples. It is found in many pnictides (compounds containing anions of Group 15) and chalcogenides (compounds containing anions of Group 16). The tetrahedral holes are filled in an ordered manner so they alternate throughout the structure. The coordination number of both ions is four.

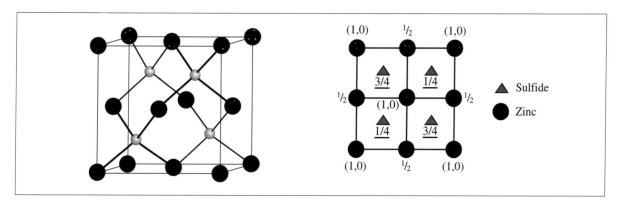

Figure 1.16 Zinc blende structure

Worked Problem

Q Show by atom counting that zinc blende contains four formula units of zinc sulfide.

A Zinc and sulfide are both tetrahedrally coordinated, and so the structure could be drawn either way round, *i.e.* it does not matter which ions form the lattice and which fill the holes. By 'atom counting' there are 4Zn and 4S per unit cell.

1.5.4 Caesium Chloride (not Close Packed)

The structure of caesium chloride is included here because, although it is not close packed, it is often confused with, and written as, body centred when it is *not*. The structure of caesium chloride is shown in Figure 1.17. The chloride ions are on the cube corners and the ion at the centre is a caesium. In Section 1.4 we saw that a body-centred cubic lattice refers to an *identical* set of points with identical atoms at the corners and at the centre of the cube. This means that the structure of caesium chloride is *not* body-centred cubic. Many alloys, such as brass (copper and zinc) possess the caesium chloride structure.

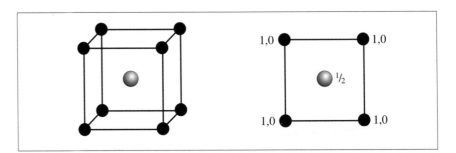

Figure 1.17 Caesium chloride structure (not close packed and not body centred)

Worked Problem

Q Use 'atom counting' to determine the number of formula units of caesium chloride per unit cell, and hence determine the lattice type.

A The lattice type is primitive (P) as there is only one Cs and one Cl in the unit cell (*i.e.* one formula unit).

1.5.5 Nickel Arsenide (Hexagonal Close Packed, Octahedral Holes Fully Occupied)

The nickel arsenide structure is shown in Figure 1.18. Arsenide ions in identical close-packed layers are stacked directly over each other, with nickel ions filling all the octahedral holes. The larger arsenide anions are in a trigonal prism of nickel cations. Both types of ion have a coordination number of six.

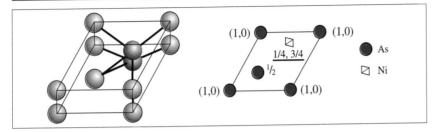

Figure 1.18 Nickel arsenide structure

1.5.6 Wurtzite (Hexagonal Close Packed, Tetrahedral Holes Half Occupied)

The wurtzite structure is the high-temperature form of zinc sulfide, where only half of the tetrahedral holes are filled in an ordered manner, such that *either* 3/8 and 7/8 *or* 5/8 and 1/8 are filled (Figure 1.19). It is noteworthy that there is no hexagonally close-packed fluorite equivalent, as Nature never simultaneously fills holes which are only one quarter of the unit cell apart.

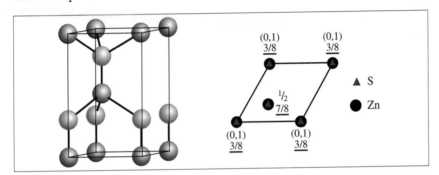

Figure 1.19 Wurtzite structure

1.5.7 Rutile (Distorted Hexagonal Close Packed, Half Octahedral Holes Occupied)

The structure typified by the rutile form of TiO_2 is shown in Figure 1.20. It has distorted TiO_6 octahedra that form columns by sharing edges, resulting in coordination numbers of six and three respectively for titanium and oxygen. The titanium ions are in a body-centred arrangement, with two oxygens in opposite quadrants of the bottom face, two oxygens directly above on the top face and two oxygens in the same plane as the central titanium inside the unit cell.

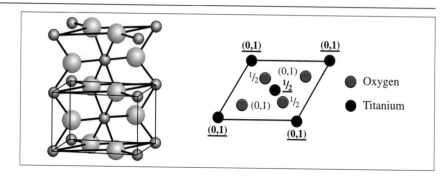

Figure 1.20 Distorted hexagonal close-packed structure of rutile

1.6   Ionic Radius Ratios and Predicting Structure

Owing to the wave nature of electrons, the ionic radius of an element is not an easily measured quantity. The numbers which appear in tables are essentially empirical values, such that when the ionic radii of an anion and cation are added together the result is the distance apart the ions would be if measured in an ionic compound. Evidence for this idea of fixed size of ions is derived from interionic distances measured by X-ray diffraction methods (Table 1.5).

Table 1.5 Differences between the sodium and potassium bond distances for the metal halides

Pair and measured interionic distance (Å)	KF	2.66	KCl	3.14	KBr	3.29	KI	3.53
	NaF	2.31	NaCl	2.81	NaBr	2.98	NaI	3.23
Difference		0.35		0.33		0.31		0.30

In Table 1.5, since the MX distance is simply the sum of the ionic radii, where

$$MX = r(M^+) + r(X^-)$$

then for KX:

$$KX = r(K^+) + r(X^-)$$

and for NaX:

$$NaX = r(Na^+) + r(X^-)$$

The difference must just be $r(K^+) - r(Na^+)$, which is approximately 0.30 Å. However, even though the *differences* can be calculated, it is not possible to calculate the *absolute* values.

One strategy to solve this problem was put forward by Pauling[4] and often forms the basis for tabulated values. Pauling argued that on theoretical grounds the size of an ion is inversely proportional to its effective nuclear charge (Z^*):

$$Z^* = Z - S$$

where Z is the atomic number and S is the effect from shielding by the inner electrons; S is determined by a number of rules formulated by Slater.

1.6.1 Predicting Structures

Previous work in this chapter has shown that a number of possible structures exist for a particular formula. For example, AB_2 could form either the rutile or fluorite structures. Although differences between the electronegativity of the ions is important, the vital factor is the relative size of the ions. This can be used to predict which structure is most likely to form.

So that energy is minimized, the cations and anions must be in contact; as many anions as possible must surround the cations (as anions are normally bigger than cations); and the geometry of the anions around the cation must minimize anion–anion repulsions.

1.6.2 Ionic Radius Ratios

The cation/anion radius ratio, r^+/r^-, provides a simple, but at best approximate, approach to predicting coordination numbers and hence structures. This uses the data from empirical tables (which treats the ions as hard spheres) to predict which is the most favoured structure. Particular ratios thus favour particular coordination numbers, as given in Table 1.6; however, the information should be treated with caution, as actual structures do not always conform to those which are predicted in this way.

Table 1.6 Calculation of radius ratios

r^+/r^-	Coordination number	r^-/r^+
1.000–0.732	8 (cube)	1.000–1.37
0.732–0.414	6 (octahedron)	1.37–2.42
0.414–0.225	4 (tetrahedral)	2.42–4.44
0.225–0.155	3 (triangle)	4.44–6.45
0.155–0	2 (linear)	6.45–

Worked Problem

Q Calculate the minimum radius ratio expected for a coordination number of eight.

A Figure 1.21. Distance AB $= a = 2r^-$. AC $= \sqrt{3}a = 2(r^+ + r^-)$

$$r^- = \frac{a}{2} \quad r^+ = \frac{\sqrt{3}a}{2} - r^-$$

$$\frac{r^+}{r^-} = \frac{\dfrac{\sqrt{3}a}{2} - \dfrac{a}{2}}{\dfrac{a}{2}} = \sqrt{3} - 1 = 0.732$$

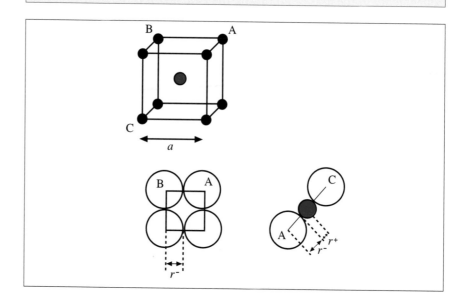

Figure 1.21 Cubic arrangement of eight anions packed around a central cation

Summary of Key Points

1. Types of Solid
Properties of different types of solid. Difference between crystalline and non-crystalline materials. Fundamental types of crystal lattice and their properties.

2. Unit Cells and Crystal Lattices
Introduction of the concept of the unit cell (fundamental repeating unit). 'Atom counting' to determine formulae.

3. *Close-packed Structures*
Generation of structure through packing atoms or ions (spheres) together so that there is the least space wasted. Building up of simple AB and AB_2 type structures by filling of tetrahedral and octahedral holes.

4. *Ionic Radius Ratio*
Use of empirical radii to predict the most likely structure from maximum/minimum possible coordination numbers of the cation.

Problems

1. On the two-dimensional grid (Figure 1.22), draw a primitive and a non-primitive unit cell.

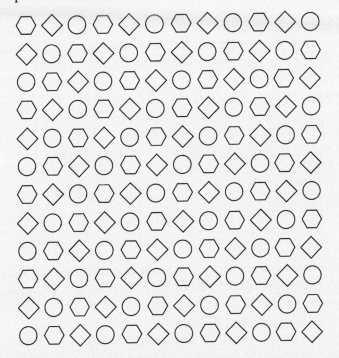

Figure 1.22 Two-dimensional array of three-component repeating lattice

2. (i) What is the formula of a compound based on a cubic close-packed structure in which one quarter of the tetrahedral holes are filled? What are the coordination numbers of both the atoms and the filled holes?

(ii) What is the formula of a compound whose structure is based on hexagonal close packing when one quarter of the octahedral holes are filled? What are the coordination numbers of both the atoms and the filled holes in this case?

3. Draw a body-centred tetragonal unit cell. Mark on the diagram the cell parameters a and c. Define the conditions expected for a tetragonal unit cell.

4. Using the projection of the unit cell for hexagonal close packing, draw four cells in a 2×2 formation. Hence, draw the outline of the hexagon which gives hexagonal close packing its name.

5. Use the counting of atoms principle to determine the ratio of octahedral holes:tetrahedral holes:atoms for hexagonal close packing.

6. β-Tungsten crystallizes with a body-centred cubic structure. Draw a projection of the unit cell of β-tungsten. Does β-tungsten crystallize with a close-packed structure? Given that the unit cell parameter for β-tungsten is 3.15 Å, calculate the ionic radius for β-tungsten.

7. Given that the crystallographic density is defined by the equation below and that the measured density of β-tungsten is 19.5 g cm^{-3}, use the data from question 6 to calculate the relative atomic mass of tungsten.

$$\text{density} = \frac{\text{mass of 1 formula unit} \times z}{\text{volume}}$$

8. HgSe is a semiconductor which can crystallize in both the zinc blende and wurtzite structures. At high temperatures the hexagonal close-packed wurtzite structure is preferred, but at lower temperatures the cubic close-packed form dominates.
(i) Describe what is meant by the underlined terms, and hence, differentiate between the two types of close-packed structures.
(ii) Draw a projection of the cubic close-packed structure of HgSe.
(iii) CeO$_2$ crystallizes with the fluorite structure at all temperatures where no hexagonal close-packed polymorph exists. By consideration of the type of hole being filled, explain why this could be predicted.

9. Lithium oxide crystallizes with the anti-fluorite structure, with lithium ions filling all the <u>tetrahedral holes</u> in a close-packed array of oxide ions.

(i) Describe the meaning of the underlined words.

(ii) Explain why the cations and anions lie in the opposite positions to those of calcium and fluoride in the fluorite structure.

(iii) What are the coordination numbers of lithium and oxygen? Using the 'atom counting' principle, determine the number of formula units per unit cell.

10. Given that the ionic radius of Fe^{2+} is 0.77 Å and that of O^{2-} is 1.26 Å, predict a likely structure for FeO using ionic radius ratios.

References

1. C. N. R. Rao and J. Gopalakrishnam, *New Directions in Solid State Chemistry*, Cambridge University Press, Cambridge, 1997.
2. A. R. West, *Basic Solid State Chemistry*, Clarendon Press, Oxford, 1997.
3. A. F. Wells, *Structural Inorganic Chemistry*, Clarendon Press, Oxford 1985.
4. L. Pauling, *J. Am. Chem. Soc.,* 1929, **51**, 1010.

2
Solid State Energetics

In Chapter 1 the differences between ionic and covalent bonding were briefly discussed. This chapter looks more closely at the principles and energetics of the ionic model, in particular how ions are assembled to form structures and what factors can be used to predict the stability of theoretical structures.

Aims

By the end of this chapter you should be able to:

- Describe the important properties of ionic materials
- Define the terms lattice energy, electron affinity and ionization energy
- Understand how to use the Madelung and Kapustinskii equations
- Use energy terms to determine unknown quantities and to predict likely structures

2.1   Assemblies of Ions and the Ionic Model

In Chapter 1 an ionic solid was defined as a lattice composed of anions and cations, where the atoms have lost or gained electrons to become ions. The loss or addition of electrons effectively completes a stable octet of outer electrons to create the ion.

The general structure of ionic solids results in the following properties:

1. **High melting points.** Ionic compounds contain strong non-directional interactions where the electrostatic attraction of

oppositely charged ions creates a network which requires a lot of energy to break down.

2. **Brittle**. Although ionic solids have high melting points, they also cleave relatively easily along certain directions. This can be observed from the flat faces seen in many mineralogical samples. A force sufficient to displace the ions slightly will map the negative ions on to negative ions and positive ions on to positive ions, *e.g.* in Figure 2.1 for sodium chloride, and the electrostatic forces which made the bonds previously so strong become repulsive and the material cleaves.

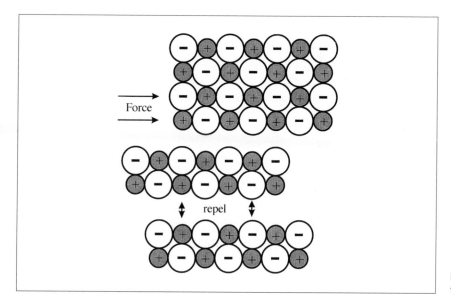

Figure 2.1 Effect of mechanical force on an ionic lattice

3. **Electrical Conductivity**. In the solid the ions are fixed on their lattice sites and electrical conductivity is poor. However, in the molten state the ions are free to move and conduct electricity in the presence of an electric field.

4. **Solubility**. Ionic compounds generally dissolve easily in polar solvents such as water. At first sight, the energy lost in breaking up the ionic lattice could be considered to be too high for dissolution; however, the solvation energy from interaction between the polar solvent and the ions, together with the increase in entropy as the lattice breaks up, compensates for the lattice energy.

5. **Coordination numbers**. Ionic compounds containing cations and anions tend to show much lower coordination numbers than metallic lattices, but higher than covalently bonded compounds. This is due to the relatively small size of the cations and relatively large size of the anions. It is not geometrically feasible to locate many large anions closely around each cation.

These properties should be treated with caution and are not definitive for ionic solids in general, since some ionic compounds do not show all these properties. For example, ammonium salts such as ammonium nitrate have low melting points (*ca.* 150–300 °C), and some compounds containing doubly charged ions such as magnesium oxide, MgO, have very low solubility.

2.1.1 Evidence for the Ionic Model

Some evidence for ionic, rather than covalent, bonding can be gathered from electron density distribution maps determined by X-ray diffraction (Chapter 3). Ions are observed as essentially spherical, highly concentrated distributions of electron density on the lattice points, with a small diffuse halo around the outside containing few electrons between the ions. This contrasts markedly with covalent bonding, where the shared electrons between the atoms generate a much higher electron density between the extremes.

Worked Problem

Q Figure 2.2 shows two schematic electron density maps for a portion of an ionic lattice and a covalent molecule. By considering each line as a contour, where closely spaced lines show changing density, which electron density map is appropriate for an ionic lattice?

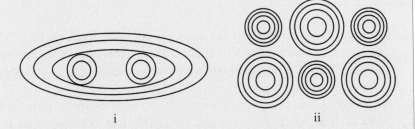

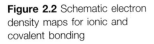
Figure 2.2 Schematic electron density maps for ionic and covalent bonding

i   ii

A Diagram (ii) is the ionic lattice, as there is very low electron density between the lattice points and the contours show an essentially spherical electronic distribution.

Such experimental evidence is not always available, so determination of a compound as fundamentally ionic is usually achieved by comparison with a model: the **ionic model**. The ionic model treats an ionic solid as an array of oppositely charged spheres held together by electrostatic (coulombic) attraction. Comparison of the thermodynamic properties of this theoretical arrangement with those determined by experiment, with a margin for experimental error, allows the compound to be classed ionic if agreement is good.

It should be noted at this stage that thermodynamic calculations are only as reliable as the experimental data placed in them. As science progresses, tabulated values for energetic terms continually change, and the most up-to-date tables should be used. Similarly, this is also true for ionic radii, which are continually being updated as the data available through use of the X-ray diffraction technique increase.

Ionization Energy

The molar first ionization energy is defined as the *change in internal energy at 0 K* for the process:

$$M_{(g)} \rightarrow M^+_{(g)} + e^-$$

in which one mole of gaseous atoms becomes one mole of gaseous monopositive ions. The second ionization energy therefore involves the loss of another electron from the already positively charged ion also at 0 K, *i.e.*

$$M^+_{(g)} \rightarrow M^{2+}_{(g)} + e^-$$

Ionization energies are often referred to as **ionization potentials** and are given the unit the eV (electronvolt). This is the energy an electron requires when it falls through a potential difference of one volt (1 eV = 96.485 kJ mol^{-1}).

Several points should be noted about ionization energies. Firstly, these reactions are *always* endothermic processes, because energy is *always* required to remove an electron from an atom. Secondly, subsequent ionization energies are always greater and progressively larger than the first. It becomes more difficult to remove the negative electrons when the atom has gained one or more positive charges. Also removal of electrons reduces inter-electron repulsion: the remaining electrons become closer to the nucleus and increasingly harder to remove. Finally, this process refers to an atom in the *gas* phase. Therefore, if the material is not gaseous and atomized, additional energy must be put into the system to generate gaseous atoms before the ionization energies can be used.

As the size of the atom increases the electrons are in general easier to remove, because the outer electrons are held less tightly as they are further away from the nucleus and shielded by the inner electrons.

Table 2.1 Ionization potential for the Group 2 elements

Element	Magnesium	Calcium	Strontium	Barium
1st I.P. (kJ mol⁻¹)	738	590	550	503
2nd I.P. (kJ mol⁻¹)	1451	1149	1064	965

Data taken from C. E. Moore, *Ionization Potential and Ionization Limits Derived from the Analyses of Optical Spectra,* National Bureau of Standards, Washington, 1970.

Therefore, the ionization energies generally decrease as a particular periodic group is descended. Table 2.1 gives the first and second ionization energies for Group 2.

Electron Affinity

The molar first **electron affinity** is defined as the energy *released at 0 K* for the process

$$X_{(g)} + e^- \rightarrow X_{(g)}^-$$

This is somewhat confusing, as normally thermodynamics uses a negative sign to indicate an exothermic reaction and here a *positive E_a* is actually an exothermic process. An element has a positive electron affinity if the incoming electron can enter an orbital where it experiences a strong attraction to the nucleus. For example, in the case of fluorine, only one electron is requires to complete a full outer shell and the electron affinity is +3.40 eV. Some electron affinities are summarized in Table 2.2.

In a similar way to the ionization energies, the sign of the electron affinity changes as more electrons are added. For the second and successive electron affinities, the negatively charged electrons are being added to an already negatively charged ion, and additional energy is

Table 2.2 Table of some elements and their electron affinities (eV)

Element	Electron affinity	Element	Electron affinity
N	−0.07	P	0.75
O	1.46	S	2.08
F	3.40	Cl	3.62
Ne	−1.2	Ar	−1.0

From H. Hotop and W. C. Lineberger, *J. Phys. Chem. Ref. Data,* 1985, **14**, 731.

required to overcome the electrostatic repulsion. The effect of the electrostatic interaction is overwhelming, and means that second, third and higher electron affinities are always endothermic, although the first electron affinity can be exothermic or endothermic.

For example, the formation of O^{2-} ions from O atoms is perhaps surprisingly endothermic, considering that under conventional terms one would expect oxygen to 'want' a complete octet of electrons:

$$O_{(g)} + e^- \rightarrow O^-_{(g)} \qquad\qquad E_a = +142 \text{ kJ mol}^{-1}$$

$$O^-_{(g)} + e^- \rightarrow O^{2-}_{(g)} \qquad\qquad E_a = -844 \text{ kJ mol}^{-1}$$

Formation of an oxide ion from a gaseous oxygen atom is an endothermic process with an overall electron affinity of -702 kJ mol^{-1}:

$$O_{(g)} + 2e^- \rightarrow O^{2-}_{(g)} \qquad\qquad E_a = -702 \text{ kJ mol}^{-1}$$

It is worth pointing out that ionization energies are always *much larger* than the corresponding electron affinities, so that for any pair of species the energy required to ionize the more electropositive element is greater than the energy obtained by adding one electron to the more electronegative element (since the first ionization energy is greater than the first electron affinity). Therefore, in the gas phase, if there is no external means of gaining energy, the elements exist as *neutral atoms and not as ions*.

2.2 Lattice Formation; Electrostatic Interaction of Ions

The forces which hold an ionic idealized lattice together are entirely electrostatic, and may be calculated by summing together all the electrostatic repulsions and attractions of cations and anions in the lattice.

The attractive force (F) between a cation of charge $+z_1e$ and an anion of charge $-z_2e$ separated by a distance r, as shown in Figure 2.3, is given by Coulomb's law:

$$F = \frac{-z_1z_2e^2}{4\pi\varepsilon_0 r^2}$$

where e is the charge on an electron and ε_0 is the permeability of free space.

The attractive potential energy (ϕ_c) of the system (the work required

Figure 2.3 Interaction of two ions separated by the distance r

to go from $r = \infty$ to some value of r) can be determined by integrating this equation over the limits $r = \infty$ to $r = r$ with respect to r, giving

$$\phi_c = \frac{-z_1 z_2 e^2}{4\pi\varepsilon_0 r}$$

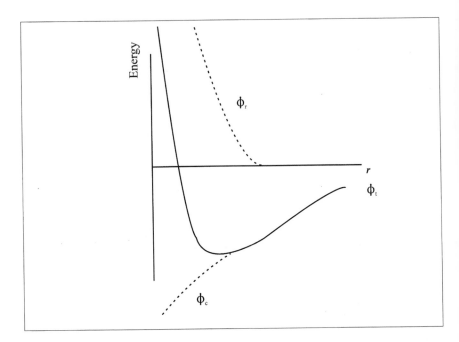

Figure 2.4 Overall attractive energy (ϕ_t) resulting from the electrostatic interaction of two ions

This expression produces the graph shown in Figure 2.4 for two ions being brought together from infinity. Clearly the ions are preventing from coalescing at $r = 0$ by some repulsive force. A repulsive force is set up owing to 'overlap' of the filled electron shells of the ions. The system comes to equilibrium where the attractive force is balanced by the repulsive force, which can be modelled using the expression:

$$\phi_r = be^{-r/\rho}$$

where b and ρ are constants, r is the distance between the two ions and ϕ_r is the repulsive term. Then:

$$\phi_t = \phi_c + \phi_r = \frac{-z_1 z_2 e^2}{4\pi\varepsilon_0 r} + be^{-r/\rho}$$

Differentiation of this expression with respect to r, where the repulsive and attractive forces are in equilibrium, produces the Born–Lande potential for one pair of ions. For one mole of ions the expression must be multiplied by Avogadro's number (N_A).

This equation now gives the energy required to generate one mole of ion pairs each separated by the equilibrium value of r (r_0) from the gaseous ions brought together from infinity. This is opposite in sign and equal in magnitude to the **lattice energy** (U_L) which is defined as the *change in internal energy at 0 K when one mole of crystals breaks down into constituent ions in the gas phase*:

$$A_mB_{n(s)} \quad \rightarrow \quad mA^{n+}_{(g)} \; + \; nB^{m-}_{(g)}$$

Using this definition, since appropriately charged ions are being separated, the value of U_L is always positive (*i.e.* breaking of an ionic lattice is always endothermic).

The **Madelung constant**, A, is also added to the equation. This is a factor which takes into account *all* the interactions between the ions, since an ion in a lattice is influenced by more than just one other single ion of opposite charge, as considered above. Fortunately, these attractions and repulsions form a mathematical series which depends on the lattice type. The values of the Madelung constant have been calculated by computer and some are shown in Table 2.3. The final expression of the Born–Lande equation then becomes:

$$U_L = \frac{A N_A z_1 z_2 e^2}{4\pi \varepsilon_0 r_0} \left[1 - \frac{\rho}{r_0} \right]$$

The equation can then be used to calculate lattice energies, providing the Madelung constant is known and the minimum distance between cation and anion can be calculated.

Table 2.3 Madelung constants for selected structures

Lattice type	Coordination numbers	Madelung constant
Sodium chloride	6:6	1.74756
Calcium fluoride	8:4	2.51939
Zinc blende	4:4	1.63806
Wurtzite	4:4	1.64132
Rutile	6:3	2.408
Caesium chloride	8:8	1.76267

Worked Problem

Q Using the Born–Lande equation, determine the lattice energy of sodium chloride given that $\rho = 34.5$ pm.

A From the table, $A = 1.74756$, the cell parameter of sodium chloride is 5.63 Å (563 pm), so the minimum distance between sodium and chloride is just 281.5 pm (half the cell edge, see Figure 1.14). The charge on both ions is one. Therefore

$$U_{L} = \frac{1.748 \times 6.022 \times 10^{23} \times 1 \times 1 \times (1.602 \times 10^{-19})^2}{4 \times 3.14 \times 8.854 \times 10^{-12} \times 281.5 \times 10^{-12}} \left[1 - \frac{34.5 \times 10^{-12}}{281.5 \times 10^{-12}} \right] \text{J mol}^{-1}$$

$$U_{L} = 863[1-0.123] = +757\text{kJ mol}^{-1}$$

Although in this example we can easily calculate the inter-ion distance in sodium chloride, the value of r_0 can also be estimated from tabulated ionic radii.

There is much confusion about the definition of lattice energy. In some textbooks it is described in terms of the formation of the lattice *from* the gaseous ions. Throughout this text the lattice energy will be described as a positive endothermic quantity which involves *the separation of ions (bond breaking)* rather than lattice formation.

The use of the Born–Lande equation requires the structure type to be known, as the Madelung constant for a specific structure is used in the equation. If the structure of the theoretical compound is not known, or the Madelung constant is not readily available, the lattice energy can be *estimated* using the Kapustinskii equation:

$$U_{L} = \frac{(1.214 \times 10^{5})Vz_1z_2}{r_1 + r_2} \left[1 - \frac{34.5}{r_1 + r_2} \right]$$

where z_1 and r_1 are the charge and radius of the first ion, z_2 nd r_2 are the radius of the second ion and V is the number of ions in the formula of the compound. This expression uses the fact that the Madelung constant is similar for compounds with the same coordination numbers. For example, sodium chloride has 6:6 coordination, and zinc blende has 4:4 coordination; in each compound both cation and anion have the same coordination number. Both sodium chloride and zinc blende have Madelung constants about 1.7, as do wurtzite and caesium chloride.

Worked Problem

Q Use the Kapustinskii equation to calculate an approximate lattice energy for calcium chloride, $CaCl_2$.

A $r^+(Ca) = 100$ pm, $r^-(Cl) = 181$ pm, $V = 3$ and $r_1 + r_2 = 100 + 181 = 281$ pm

$$U_L = \frac{(1.214 \times 10^5) \times 3 \times 2 \times 1}{281}\left[1 - \frac{34.5}{281}\right]$$

$$= 2270 \text{ kJ mol}^{-1}$$

2.3 Born–Haber Cycles

Since energy is always conserved, the application of the First Law of Thermodynamics means that the change in enthalpy for a process does not depend on the way the energy change is carried out (this is often quoted as Hess's law). In 1919, Born[1] and Haber[2] working independently applied the First Law of Thermodynamics to the formation of ionic solids from their elements (*e.g.* Figure 2.5, for a simple binary ionic solid, A^+B^-). To calculate an unknown quantity is therefore just a matter of following the arrows in the cycle, provided all other values are known and the correct *sign* is used for each value (+ for *endothermic* processes and – for *exothermic*).

It should be noted that thermochemical cycles are often calculated at 298 K, whereas the energy terms such as electron affinity or ionization energy are defined at 0 K. Therefore, the values calculated by thermochemical cycles have an error of approximately 2–5 kJ mol^{-1}.

In Figure 2.5, each side of the cycle can be considered, so for energy to be conserved

$$-A + B + C + D = F - E$$

or

$$-A + B + C + D + E - F = 0$$

In addition to the lattice energy, electron affinity and ionization energy which have already been defined in this chapter, Born–Haber cycles also contain other quantities which allow for the fact, for example, that metals are not in the gaseous state at 298 K and that the halogens do not exist as mononuclear species.

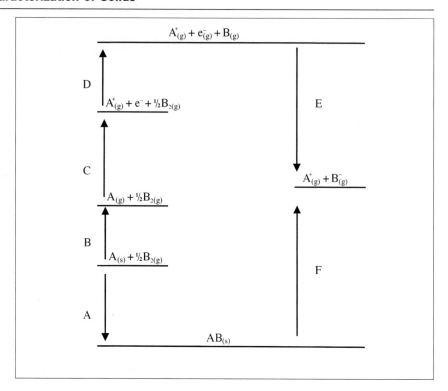

Figure 2.5 Born–Haber cycle for the formation for a Group 1 halide

2.3.1 Enthalpy of Formation, ΔH_f

The enthalpy change required to produce one mole of any compound from its constituent elements in their standard state

$$M_{(s)} + \tfrac{1}{2}X_{2(g)} \rightarrow MX_{(s)}$$

is called the **enthalpy of formation** for that compound (A). From Figure 2.5, the first stage of the cycle is formation of the elements in their standard state *from the ionic lattice, i.e. $-\Delta H_f$.*

A general method is being introduced here, in which enthalpies rather than free energies are being considered. However, the energetics of lattice formation are governed by the equation for Gibbs free energy, *i.e.*

$$\Delta G = \Delta H - T\Delta S$$

where ΔG is the Gibbs free energy change, ΔH is the enthalpy change and ΔS is the entropy change during the reaction. A negative value of ΔG (energy given out during the process) indicates a lower energy state for the system after the reaction and hence reaction is favoured.

The formation of a lattice from its gaseous ions is, however, very exothermic (ΔH is negative) so the loss in entropy is of negligible importance. The thermodynamic properties of crystal lattices therefore often focus on changes in enthalpies rather than in free energies.

2.3.2 Enthalpy of Sublimation, ΔH_{sub}

The formation of one mole of gaseous atoms from one mole of the atoms in the standard state of the element

$$M_{(s)} \rightarrow M_{(g)}$$

is called the enthalpy of sublimation, as in stage B.

2.3.3 Dissociation Enthalpy, ΔH_{diss}

In many cases, element B will exist as a diatomic molecule in its standard state (*e.g.* the halogens). To include it in the cycle the element must be dissociated to form monoatomic atoms, as in stage C:

$$X_{2(g)} \rightarrow 2X_{(g)}$$

Since we only have a singular X atom in the cycle we only need *one half of the dissociation enthalpy* in order to have one mole of X atoms. This value, for an element such as chlorine which normally exists as a diatomic molecule, is often known as the enthalpy of atomization, ΔH_{at}. This process is represented as:

$$\tfrac{1}{2} X_{2(g)} \rightarrow X_{(g)}$$

If X is a Group 16 element (chalcogenide), *e.g.* sulfur, instead of a halogen, then the vaporization of the solid sulfur to the gaseous sulfur would have to be included in the cycle instead of ΔH_{diss}.

Stage D is the change in internal energy for the electron transfer to the non-metal element ($-E_a$) and stage E is the lattice enthalpy. The lattice energy (U_L) and the lattice enthalpy (H_L) vary by ~5 kJ mol^{-1} due to the calculation being carried out at 298 K rather than 0 K. The difference is so small compared with the magnitude of the quantity that the two terms are often interchangeably.

This whole cycle can be summarized as:

$$\Delta H_L = \Delta H(\text{atomization}) + \Delta H(\text{ionization}) - \Delta H_f$$

If any five of the quantities A–F are known, the sixth can be calculated.

Worked Problem

Q Use the data below to construct a thermochemical cycle for the formation of potassium chloride from its elements and hence calculate the lattice energy of potassium chloride.

$$K_{(s)} \rightarrow K_{(g)} \qquad\qquad +89 \text{ kJ mol}^{-1}$$

$$\tfrac{1}{2}Cl_{2(g)} \rightarrow Cl_{(g)} \qquad\qquad +122 \text{ kJ mol}^{-1}$$

$$Cl_{(g)} + e^{-} \rightarrow Cl^{-}_{(g)} \qquad\qquad -355 \text{ kJ mol}^{-1}$$

$$K_{(s)} + \tfrac{1}{2}Cl_{2\,(g)} \rightarrow KCl_{(s)} \qquad -438 \text{ kJ mol}^{-1}$$

A See Figure 2.6.

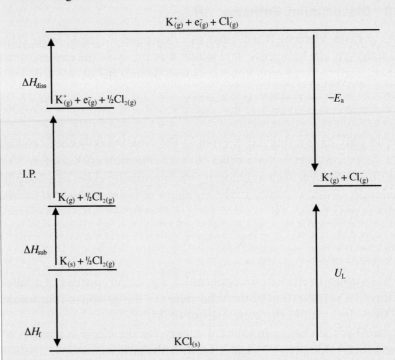

Figure 2.6 Born–Haber cycle for the formation of KCl

$$K_{(g)} \rightarrow K^{+}_{(g)} + e^{-} \qquad +425 \text{ kJ mol}^{-1}$$

$$U_L = 89 + 425 + 244 - 355 + 438 = +719 \text{ kJ mol}^{-1}$$

This value is in good agreement with the experimental lattice energy of 709 kJ mol^{-1}, which strongly suggests that the ionic model is valid for potassium chloride.

Some caution should be applied when using data of this kind to predict the stability of compounds. The routes under which those compounds can decompose should be considered. For example, calculation of a lattice energy for the theoretical compound $CaCl_{(s)}$ produces an exothermic enthalpy of formation of approximately -200 kJ mol^{-1}. The compound CaCl ought therefore to exist. Chemical knowledge dictates that this compound does not exist. The reason becomes apparent by considering the equation:

$$2CaCl_{(s)} \rightarrow CaCl_{2(g)} + Ca$$

This reaction is more exothermic, with an enthalpy change of more than -600 kJ mol^{-1}. Any CaCl which might be formed disproportionates: the Ca^+ species become Ca^{2+} and Ca^0.

2.4 Bond Enthalpies and Stability

In understanding the chemistry of the main group elements, an appreciation of the factors which affect the stability of compounds is paramount. Just as lattice energies can be used to predict stabilities of ionic lattices, bond enthalpies can be used to explain wide-ranging variations in the properties of the main group elements. For example, the differences in the physical and chemical properties of the dioxides of carbon and silicon can be explained by considering bond enthalpy terms. Similarly, why CH_4 is a very stable compound while PbH_4 is extremely unstable can be explained. In addition, other trends which occur in the chemistry of the main group elements can be understood in terms of simple energetics.

2.4.1 Stability and Instability; Kinetic and Thermodynamic Stability

In order to define the stability of an inorganic material we need to be precise about what we mean by stability. For example, $CaCl_2$ could conceivably decompose to a range of known or hypothetical compounds, e.g. to $Ca + Cl_2$ or to $CaCl + \frac{1}{2}Cl_2$, or to $CaCl_3 + 2Ca$. All of these routes would involve a large increase in the free energy of the system. We can therefore describe $CaCl_2$ as being *thermodynamically stable* with respect to other calcium chlorides/calcium/chlorine. A compound may be thermodynamically unstable, but *kinetically stable, i.e.* it has a negative free energy with respect to all possible decomposition products. For example, a sample of NO in a sealed container at room temperature does not decompose. However, the free energy change for the reaction

$$2NO \rightarrow \frac{1}{2}N_2 + \frac{1}{2}O_2$$

is -87 kJ mol^{-1}, indicating that NO is thermally unstable with respect to its constituent elements. Therefore, NO can be referred to as being *kinetically stable*; it is unstable in terms of energy, but the rate at which the change occurs at room temperature is so slow as to be undetectable. However, if NO is exposed to air, reaction with oxygen takes place to produce NO_2. NO is therefore not stable in air, *i.e.* with respect to oxidation.

In conclusion, when referring to the stability of a compound one must be *precise*. Many inorganic compounds can be prepared which are thermodynamically unstable but kinetically stable with respect to decomposition along various routes. In addition, inorganic compounds which react with oxygen or water can often be handled in inert atmospheres, where they can be regarded as perfectly stable.

2.5   Bond Dissociation Energies and Bond Enthalpy Terms

Before any discussion about the application of bond enthalpy terms, it is important to check the definitions of bond dissociation energies and bond enthalpy terms.

The bond dissociation energy is defined as the enthalpy change associated with the reaction in which one mole of bond is *homolytically* broken, reactants and products being in the ideal gas state at 1 atm and 298 K.

Thus for methane, the bond dissociation energy for the first C–H bond is the energy change for the process:

$$CH_{4(g)} \rightarrow CH_{3(g)} + H_{(g)} \qquad\qquad \Delta H = +432 \text{ kJ mol}^{-1}$$

This bond dissociation energy is specific to the reaction above, *i.e.* the energy for this C–H bond-breaking process depends on factors such as the other atoms attached to the carbon atom and the geometry of CH_4 and CH_3. For example, if the remaining C–H bonds in methane are broken:

$$CH_{3(g)} \rightarrow CH_{2(g)} + H_{(g)} \qquad\qquad \Delta H = +470 \text{ kJ mol}^{-1}$$

$$CH_{2(g)} \rightarrow CH_{(g)} + H_{(g)} \qquad\qquad \Delta H = +416 \text{ kJ mol}^{-1}$$

$$CH_{(g)} \rightarrow C_{(g)} + H_{(g)} \qquad\qquad \Delta H = +335 \text{ kJ mol}^{-1}$$

While bond dissociation energies are important in terms of the general understanding of physical chemistry, they are of little general use in predicting the stability of inorganic compounds, as they cannot be transferred between compounds.

Bond enthalpy terms, on the other hand, are quantities assigned to *each bond* in a molecule such that the sum over all bonds is equal to the enthalpy change associated with the conversion of the molecule into separate atoms. Bond enthalpy terms are assumed to be *constant*, and therefore transferable from molecule to molecule. So for methane:

$$CH_{4(g)} \rightarrow C_{(g)} + 4H_{(g)} \qquad\qquad \Delta H = 1663 \text{ kJ mol}^{-1}$$

so that the bond enthalpy term $B(C–H)$ is $\frac{1}{4}\Delta H = 416$ kJ mol^{-1}, *i.e.* the *average* for the four C–H bonds in CH_4.

Bond enthalpy terms are not precise, particularly where there are changes in bond valency. For example, the value for C=C is +612 kJ mol^{-1}, which is considerably less than twice the value for C–C. The values for $B(C–H)$ and $B(C–C)$, typically given in the literature as +413 kJ mol^{-1} and 347 kJ mol^{-1}, respectively, are good for working out the enthalpies of formation of a wide range of alkane hydrocarbons. This is because carbon always forms four covalent tetrahedral bonds in these compounds. For many inorganic systems, the valency of a particular element can vary markedly, and with it the number of bonds and the geometrical distribution of those bonds. For example, consider ClF_3 and ClF. The bond enthalpy terms for the Cl–F bond, derived from the enthalpies of formation of these compounds, are very different at +174 and +255 kJ mol^{-1}, respectively. This demonstrates that transporting a chlorine–fluorine bond enthalpy term from one oxidation state of chlorine to another would produce considerable errors. Therefore, bond enthalpy terms are a useful guide, but their use should be treated with caution.

Some homolytic (between the same element) and heterolytic (between different elements) bond enthalpies are summarized in Table 2.4.

2.5.1 Variations in Bond Enthalpy Terms down a Main Group

Some general trends in bond enthalpies down a main group can be noted from Table 2.4. In Groups 14, 15, 16 and 17, if the element in the period from Li to F is A_1, the period Na to Cl is A_2, *etc.*, then

1. In any vertical group A_1, A_2, *etc.*, the bond enthalpy $B(A–X)$ diminishes down the group (A_1 to A_n) provided there are no lone pairs on X (*e.g.* H, C, but *not* N, O or F). So, for example, we have the data in Table 2.4 for the hydrides (A–H).
2. When there are lone pairs on X, the bond enthalpy order is usually:

$$B(A_1–X) < B(A_2–X) > B(A_3–X) > B(A_4–X)$$

Table 2.4 Selected homolytic and heterolytic bond enthalpies (kJ mol^{-1})$^{3-7}$

Homolytic bond enthalpies						
H–H						
432						
Li–Li	Be–Be	B–B	C–C	N–N	O–O	F–F
205	208	293	346	167	142	158
Na–Na	Mg–Mg	Al–Al	Si–Si	P–P	S–S	Cl–Cl
72	129	183	222	201	226	249
K–K	Ca–Ca	Ga–Ga	Ge–Ge	As–As	Se–Se	Br–Br
49	105	115	188	146	172	190
Rb–Rb	Sr–Sr	In–In	Sn–Sn	Sb–Sb	Te–Te	I–I
45	84	100	146	121	126	149

	Heterolytic bond enthalpies					
	A–H	A–F	A–Cl	A–Br	A–O	A=O
C	411	485	346	290	358	736·
Si	318	565	381	310	466	638
Ge	285	452	349	276	385	–
N	386	313	283	–	201	607
P	322	490	322	263	335	544
As	247	406	309	256	301	389
Sb	–	402	314	264	–	–
O	459	190	218	201	144	498
S	364	284	255	217	468	–
Se	276	303	251	201	–	–
F	565	158	249	248	190	–
Cl	428	249	240	216	218	–
Br	362	248	216	190	201	–
I	295	278	210	177	201	–

Trend (1) is reasonably expected. As we proceed down a group, for example from carbon to lead, the atoms get larger and the orbitals involved in the bond have a more diffuse electron cloud, a poorer overlap with orbitals of X and thus form weaker bonds.

Trend (2) is more complex, but two factors contribute to inversion of the stability of A$_1$–X and A$_2$–X. One factor is repulsion between non-bonding electrons. The F–F bond is much weaker than the Cl–Cl bond for this reason. The non-bonding pairs for fluorine lie much closer to the nucleus and thus closer to the non-bonding pairs on the other nucleus, causing more repulsion and weakening the F–F bond relative to the Cl–Cl bond. The second factor is the possibility of d-orbital involvement in π-bonding for second-row elements. The second-row elements have vacant low-lying d-orbitals which can form a dπ–pπ interaction with p-orbitals on X, which are normally filled. This acts to strengthen the bond from the second-row element to X (where X has lone pairs), and helps account for trend (2). Such $d\pi$–$p\pi$ bonding can occur for heavier elements in a group, but since the orbitals are larger and the interaction is spread out over a larger volume, it will be a weaker effect.

Application of trends (1) and (2) can explain some obvious differences between chemical reactivity and structure in the main groups.

Double and Single Bonds

A major difference in the chemistries of first- and second-row elements is apparent in the relative stability of compounds with double and single bonds. Carbon–carbon double bonds are well known, whereas only recently have compounds with silicon–silicon double bonds been synthesized. Such compounds readily react to form other compounds in which the double bond is broken.

Worked Problem

Q Figure 2.7 shows the formation of four M–O single bonds from two M=O double bonds, where M = Si and C. Consider the enthalpy change for this reaction, and hence predict the preferred form of the oxide for each of Si and C.

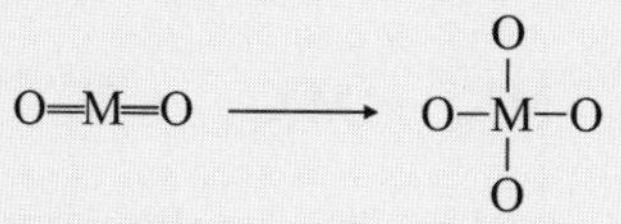

$$O=M=O \longrightarrow \begin{matrix} O \\ | \\ O-M-O \\ | \\ O \end{matrix}$$

Figure 2.7 Formation of four single M–O bonds from the decomposition of two double M=O bonds

A Using the data in the Table 2.4, we find:
For carbon $(736 \times 2 - 358 \times 4) = +40$ kJ mol^{-1}
For silicon $(638 \times 2 - 466 \times 4) = -588$ kJ mol^{-1}

For carbon the reaction to form four single bonds from two double bonds would be endothermic (*i.e.* unfavourable) and hence CO_2 exists as discrete molecules. Silicon much prefers the structure with four single bonds and SiO_2 is an infinite lattice of vertex-linked SiO_4 tetrahedra. The reasons for the different structures can be rationalized by considering the bonding in them. C=O bonds are strong, presumably owing to the good $p\pi$–$p\pi$ overlap which comes from the similar sizes of the oxygen and carbon p-orbitals and the short C–O distance. For Si=O, the overlap will be much poorer with a longer Si–O distance and 3p–2p orbitals involved. In contrast, the Si–O single bond is stronger than the C–O single bond owing to $d\pi$–$p\pi$ interactions.

Stability of Gaseous Halides AXn and the Inert Pair Effect

Among the heavy post-transition metals there is a definite reluctance to exhibit the highest possible oxidation state. For example, boron is always trivalent, but thallium shows significant chemistry of the +1 oxidation state, leaving a pair of electrons coordinatively inert. This is known as the inert pair effect.

The reasons for the differences can be explained using bond enthalpies. As we have seen, the values of the first ionization energies fall as a group is descended. Bond enthalpies to a particular element also decrease.

For example, Table 2.5 contains the first four ionization energies for the Group 14 elements. There are only small differences between the ionization energies for silicon, which primarily forms compounds in the +4 oxidation state, and lead which forms both divalent (+2) and tetravalent (+4) compounds.

A better understanding of the occurrence of the inert pair effect can be observed by considering the tendency for the following reaction to proceed either in the solid or gas phase:

$$MX_n \rightarrow MX_{n-2} + X_2$$

Consider the values, in kJ mol^{-1}, for the gas phase reaction in Table 2.6. Although all these values for the decomposition are endothermic, the trend to decreasing stability of the high-valent (+4) fluoride compared with the low-valent (+2) iodide can be clearly seen. This is due to longer bonds and more diffuse orbitals giving poorer overlap and weaker bonds. Hence, as a group is descended the bonds to X become weaker,

Table 2.5 Ionization energies for the Group 14 elements (kJ mol^{-1})

	1st	2nd	3rd	4th
C	1090	2350	4610	6220
Si	786	1580	3230	4360
Ge	762	1540	3300	4390
Sn	707	1410	2940	3930
Pb	716	1450	3080	4080

Table 2.6 Bond enthalpies for the Group 14 halides (kJ mol^{-1})

Element	Fluoride	Chloride	Bromide	Iodide
C	467	346	290	213
Si	565	381	310	234
Ge	452	349	276	211
Sn	414	323	272	205
Pb	331	243	201	142

and their formation does not compensate for the endothermic energy terms such as breaking the X–X bond and the electron promotion (hybridization) energies $s^2p \rightarrow (sp^2)^3$. It is also worth noting that, as more bonds form around an atom, each successive bond normally gets weaker. For example, the series:

	ClF	ClF$_3$	ClF$_5$	
B(Cl–F)	+255	+174	+151	kJ mol^{-1}

So the weakening bonds as the group is descended is really the origin of the tendency of an element to form the $n - 2$ oxidation state.

Summary of Key Points

1. Structures formed from ions have specific properties which are related to their geometrical structure, such as high melting point, brittle nature and localized electrons.

2. The actual assembly of positive and negative ions revealed by the X-ray diffraction technique shows sound agreement with theoretical mathematical expressions, in which ions are considered point charges. This allows the stability of purely theoretical ionic structures, *e.g.* CaCl or CaCl$_3$, to be predicted.

3. Energy terms such as ionization energies and lattice energies (defined at 0 K) can be used in Born–Haber cycles to estimate unknown quantities under standard conditions.

4. Bond enthalpy terms can be used to explain differences in the reactivity and bond strengths of the main group elements and to explain general trends such as the inert pair effect and the tendency to form double or single bonds.

Further Reading

D. A. Johnson, *Some Thermodynamic Aspects of Inorganic Chemistry*, Cambridge University Press, Cambridge, 1981.

W. E. Dascent, *Inorganic Energetics*, Penguin, London, 1970.

J. E. Huheey, E. A. Keiter and R. L. Keiter, *Inorganic Chemistry*, 4th edn., Harper/Collins, New York, 1994.

N. C. Norman, *Periodicity and the s and p Block Elements*, Oxford University Press, Oxford, 1999.

Problems

1. (i) Given that the cell parameter for calcium oxide, which crystallizes with the sodium chloride structure, is 483 pm, calculate the lattice energy for CaO.

(ii) Using the following data and the lattice energy calculated in part (i), construct a thermochemical cycle for the formation of calcium oxide from its elements and hence calculate the enthalpy of formation of $CaO_{(s)}$:

ΔH_{sub} = +177; ionization potentials: first = +590; second = +1100; ΔH_{diss} = +498; first E_a = 142; second E_a = –844 (kJ mol^{-1})

(iii) Why might you expect the answer calculated in (ii) to have an error of about 5 kJ mol^{-1}?

(iv) Why is the second ionization potential for calcium much greater than the first despite forming a stable closed-shell configuration?

2. Given that strontium oxide crystallizes with the same structure as calcium oxide, use data tables to estimate the minimum cation/anion distance and hence determine the lattice energy for SrO.

3. The compound FrF cannot be isolated owing to short half-life of Fr. Use the Kapustinskii equation to calculate an approximate lattice energy for FrF, given that the sum of the francium and fluorine ionic radii is 4.2 Å

4. The experimentally determined lattice energy for rubidium chloride is 702 kJ mol^{-1}. Use the Born–Lande model to calculate the lattice energy for RbCl, and hence confirm that RbCl fits the ionic model.

5. (i) A thermochemical cycle for the formation of the aluminium sub-halides is given in Figure 2.8. Write an expression for ΔH_{dec} in terms of the other *quantities in the cycle*.

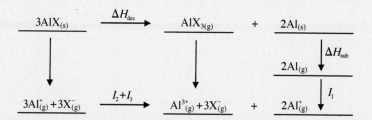

Figure 2.8 Thermochemical cycle for the formation of aluminium sub-halides

(ii) Assuming no entropy change, predict which oxidation state of aluminium is energetically favoured for the two halides, given the following data (kJ mol^{-1}):

$I_1 = +577$ $I_2 = +1816$ $I_3 = +2743$ $\Delta H_{sub} = +324$

$U(AlF) = +910$ $U(AlI) = +696$

$U(AlF_3) = +6380$ $U(AlI_3) = +4706$

6. Explain the following observations.
(i) The C–O single bond is weaker than the Si–O single bond whereas the double bonds show the reverse order of strength.
(ii) The C–C bond and the F–F bonds are both considerably weaker than the C–F single bond.
(iii) The N–N single bond is weaker than the P–P single bond.
(iv) The Si–Cl bond is easily hydrolysed although the C–F bond is resistant to attack by water.

References

1. M. Born, *Verh. Dtsch. Phys. Ges.*, 1919, **21**, 13.
2. F. Haber, *Verh. Dtsch. Phys. Ges.*, 1919, **21**, 750.
3. T. L. Cottrell, *The Strength of Chemical Bonds*, Butterworth, London, 1958.
4. U. S. National Bureau of Standards, NSRDS-NBS 31, 1970.
5. L. Brewer and E. Brackett, *Chem. Rev.*, 1961, **61**, 425.
6. L. Brewer, *Chem. Rev.*, 1963, **63**, 111.
7. R. C. Feber, Los Alamos Report, LA-3164, 1965.

3
Characterization of Solids

The total characterization of a new material is an essential part of any investigation, not only the chemical and physical properties, such as its reactivity or its magnetic properties, but also its structure. Typically, the structure of a solid material governs its observed properties. Copying and modelling of these structural features can lead to the development of new materials; for example, this procedure led to new and improved superconducting materials.

A new functional material not only has to fulfill the desired task, *e.g.* as a pigment or an ion exchanger, but it also has to be thoroughly investigated for stability and toxicity. Since cost effectiveness is particularly important for industrial materials, the preparative conditions must also be optimized to reduce the cost of starting materials and synthetic methods.

Aims

This chapter aims to outline the essential techniques for characterizing materials. By the end of this chapter you should be able to:

- Understand the principles of the diffraction technique and index cubic materials
- Understand the differences between applying spectroscopic methods to the solid state rather than molecular materials
- Interpret information from thermal techniques for determining water content and phase changes

3.1   X-ray Diffraction

X-ray diffraction is the most widely used and least ambiguous method for the precise determination of the positions of atoms in molecules and solids. Spectroscopic information (NMR, IR and mass spectrometry), which often suffices for organic chemistry, does not give such a complete picture for inorganic materials, owing to much greater structural diversity in terms of coordination numbers and geometries. Particularly in inorganic chemistry, the distances derived from the structure give additional information on bonding within and between molecules. For example, particular bond lengths and angles are indicative of certain oxidation states.[1]

3.1.1 Principles of X-ray Diffraction

X-rays interact with electrons in matter. When a beam of X-rays impinges on a material it is scattered in various directions by the electron clouds of the atoms. If the wavelength of the X-rays is comparable to the separation between the atoms, then interference can occur. For an ordered array of scattering centres (such as atoms or ions in a crystalline solid), this can give rise to interference maxima and minima. The wavelengths of X-rays used in X-ray diffraction experiments therefore typically lie between 0.6 and 1.9 Å.

Generation of X-rays

A beam of electrons striking a metal target will eject electrons ('core electrons') from the energy levels close to the nucleus of some of the metal atoms (providing the beam is higher in energy than the energy needed to remove such electrons). Once vacancies have been created, electrons from higher energy levels fall down (decay) into these orbitals. The difference in energy between these higher and lower energy levels is then emitted as an X-ray of precise energy ($\Delta E = h\nu$, where ΔE is the energy difference, ν is the frequency of the emitted radiation and h is Planck's constant).

The X-ray Tube

A schematic diagram of a typical X-ray tube is shown in Figure 3.1. The electrons are created by heating a tungsten filament in a vacuum (thermionic emission) and then accelerated by a high voltage (typically 30,000 V) towards a metal target. Core electrons are knocked out of the metal target and the X-rays, characteristic of the metal target, are produced by decay. By making a beryllium window in the tube, because beryllium has a low atomic number and is transparent to X-rays, the X-ray beam escapes from the tube.

A spectrum of the output from an X-ray tube is shown in Figure 3.2. The background is known as Bremstrahlung (or braking) radiation,

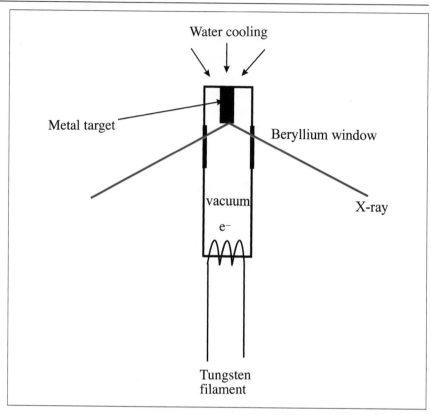

Figure 3.1 Schematic diagram of an X-ray tube.

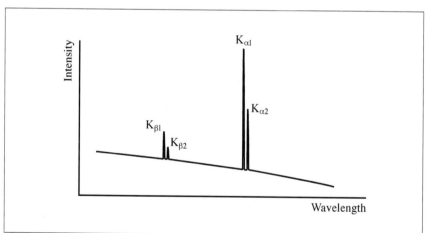

Figure 3.2 Output from an X-ray tube

which is caused by the energy loss of the X-rays as they enter the metal. This is often called 'white radiation' as it is emitted over the full wavelength range. On top of the braking radiation are sharp lines generated by the **quantized** transitions (see above). The labels describe which orbital the decaying electron has come from, the orbital being filled and the spin state of the electron.

X-ray Experiment

A single wavelength must be selected from the output from the tube to carry out the X-ray experiment. Simple filtering can get rid of some of the unwanted X-rays; for example, by using a film consisting of a metal with atomic number one below that of the metal target, the K_β lines and the white radiation can be removed. However, to obtain a single wavelength a single crystal monochromator is used. This utilizes Braggs Law (see Section 3.1.2) to select a single wavelength. Generally the α_1 line is selected, if possible, as it has the greatest intensity.

Once a single wavelength has been selected by passing the beam through the single crystal monochromator it is directed towards the sample using a collimator. The scattered X-rays are then analysed using a detector.

Miller Indices

X-rays interact with planes of atoms in the three-dimensional lattices which show the translational symmetry of the structure. Each plane is a representative member of a parallel set of equally spaced planes, and each lattice point must lie on one of the planes.

The labels used for describing these planes are known as Miller indices and are given the descriptions h, k and l, where h,k,l *take values of positive or negative integers or zero.* Consider the two planes shown in Figure 3.3. The Miller indices of these two members of a family of planes are given by the reciprocals of the fractional intercepts h,k,l along each of the unit cell directions, *i.e.* cutting at a/h, b/k and c/l. So the 2,8,1 plane would cut one half of the way along a, one eighth of the way along b and all the way along c. The parallel plane on the diagram has the same Miller indices as it stretches across two unit cells, this time cutting the axes at a, $b/4$ and $2c$, respectively. For planes which are parallel to one of the unit cell directions the intercept is at infinity, and therefore the

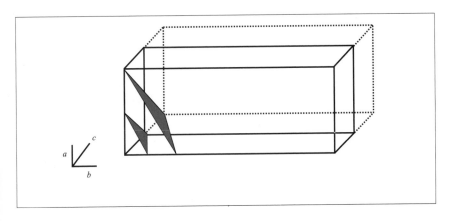

Figure 3.3 Representation of the 281 family of planes

Miller index for this axis is zero. The separation of the planes is known as the *d*-spacing and is normally denoted d_{hkl}. From the diagram, it is apparent that this is also the perpendicular distance from the origin to the plane.

Worked Problem

Q Draw a tetragonal unit cell and mark on the diagram the 231 plane and the distance d_{231}.

A See Figure 3.4. The 231 plane cuts the unit cell axes at $a/2$, $b/3$ and c, and the distance d_{231} is the perpendicular distance from the origin to the plane. Note that, although in a tetragonal unit cell the a and b axes are equivalent in length, they should be distinguished when drawing planes.

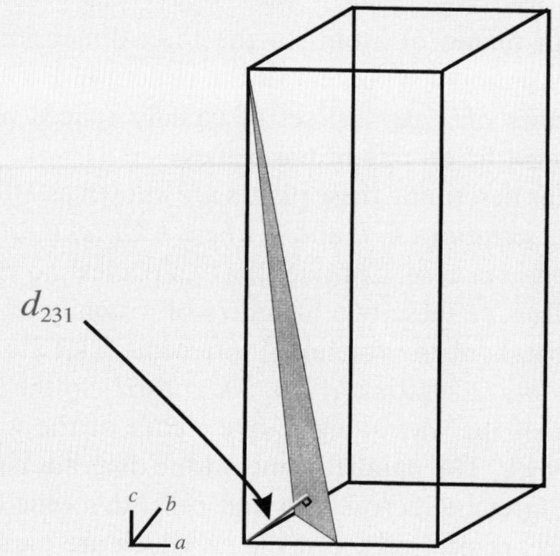

Figure 3.4 Tetragonal unit cell showing the 231 plane and the *d*-spacing

3.1.2 Scattering from Crystalline Solids and the Bragg Equation

The scattering of X-rays from a set of planes defined by the Miller indices *h,k,l* is shown in Figure 3.5. In order to observe useful data from the X-ray experiment the scattered X-ray beam from the points X and Z must produce diffracted beams which are in phase. This is only possible if the extra distance travelled by the X-ray photon from W to X and X to Y is an *integral number of wavelengths*. The path difference is dependent on both the lattice spacing d_{hkl} and the angle of incidence of the X-ray beam, θ:

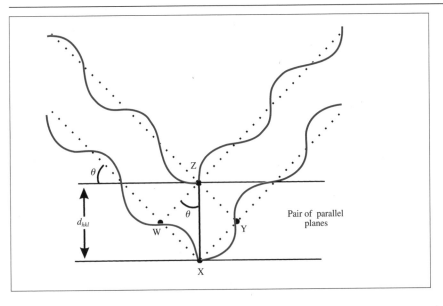

Figure 3.5 Scattering of X-rays from a parallel set of planes

$$\text{Path difference} \;=\; \text{WX} \;+\; \text{XY} \;=\; 2\,d_{hkl}\sin\theta \;=\; n\lambda$$

where n is an integer, and λ is the X-ray wavelength. This equation is the Bragg equation. The n is normally dropped (*i.e.* treated as 1), as higher order ($n = 2$, *etc.*) diffraction maxima are just equivalent to diffraction from the $n = 1$ set but at shorter d spacing.

Worked Problem

Q Second-order scattering from the 100 plane is equivalent to first-order scattering from which plane?

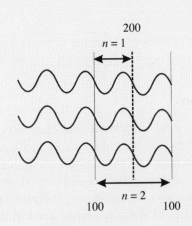

A Using the Bragg equation:

$$n\lambda \;=\; 2\,d_{hkl}\sin\theta$$

See Figure 3.6. The plane must be the 200.

Figure 3.6 First- and second-order diffraction

In any crystalline material an infinite set of planes exists with different Miller indices, and each set of planes has a particular separation. For any set of planes there will be a diffraction maximum at a particular angle, θ. By combining the equation relating d_{hkl} to the lattice parameter for a particular system with the Bragg equation, a direct relationship between the diffraction angle and the lattice parameters can be derived.

For example, for a cubic system, combining the geometric equation for a cubic system (where a is the lattice parameter):

$$\frac{1}{d^2} = \frac{h^2 + k^2 + l^2}{a^2}$$

with the Bragg equation

$$n\lambda = 2d\sin\theta$$

where $n = 1$ and rearranging for d gives

$$\frac{1}{d} = \frac{2\sin\theta}{\lambda}$$

which means

$$\frac{1}{d^2} = \frac{4\sin^2\theta}{\lambda^2}$$

Substituting for $1/d^2$ in the first equation and rearranging gives

$$\sin^2\theta = \frac{\lambda^2}{4a^2}(h^2 + k^2 + l^2)$$

By using this equation, *structural* information about the crystals under study can be obtained.

3.1.3 Single Crystal Diffraction

By far the dominant technique in solid state/materials chemistry is powder diffraction since single crystals are often difficult to synthesize and not representative of the bulk, which can show non-stoichiometry and disorder. However, for definitive characterization of physical properties such as magnetic ordering or electron transport, single crystals are often used, as the directional properties of these processes are lost in a polycrystalline sample.

The single crystal diffraction technique is covered well in many other texts[2,3] and will not be discussed in detail here. The preparation of single crystals from solid state materials is rather different from the preparation of single crystals from molecular systems and is therefore described in Chapter 4.

3.1.4 Powder Diffraction

In a single crystal experiment the alignment of the crystal, source and detector have to be correct to observe any reflections, as scattering from one particular plane at a time is measured by aligning the crystal appropriately. In contrast, a powdered sample contains an enormous number of very small crystallites, typically 10^{-7}–10^{-4} m in size, which randomly adopt the whole range of possible orientations. Therefore, when an X-ray beam strikes a powdered (often termed polycrystalline) sample, it is diffracted in all possible directions (as governed by the Bragg equation) simultaneously. Each lattice spacing in the crystal gives rise to a cone of diffraction, as shown in Figure 3.7.

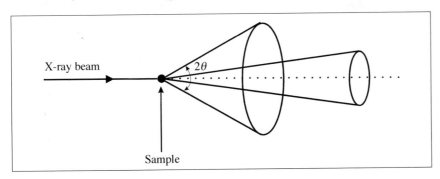

Figure 3.7 Continuous diffraction cone derived from scattering from a polycrystalline sample

Each cone is a set of closely spaced dots, where each dot represents diffraction from a single crystallite within the sample. With a large number of crystallites, these join together to form a continuous cone.

Experimental Methods

To analyse the data, the positions of the cones need to be measured. This can be achieved by using photographic film or, more often than not, by a radiation detector (diffractometer). The use of film and camera has become very limited as diffractometer technology has advanced. Although camera methods are fast to employ and require very little sample, the superior resolution and electronic storage capability of the diffractometer is preferred. For example, since each cone of dots represents diffraction from a plane of atoms, when the structure become complex the lines become too numerous and overlap, making it impossible to discern from camera methods alone which line applies to which set of atoms.

A schematic diagram of a powder diffractometer is shown in Figure 3.8. The X-rays produced by the X-ray tube are aligned to fall on the sample through a slit and are scattered in all directions. By scanning the detector around the sample along the circumference of a circle, it is made to cut through the diffraction cones at various diffraction maxima. The X-ray diffraction pattern (Figure 3.9) displays intensity as a function of the detector angle, 2θ. The reflection geometry of powder diffractometers enables slits close to the detector to remove noise and leads to well-resolved data, where the sample behaves like a mirror which focuses the beam on to the detector.

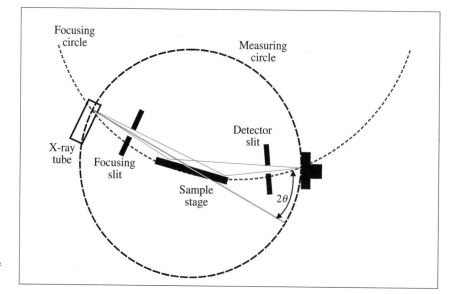

Figure 3.8 Schematic diagram of a powder X-ray diffractometer

Powder X-ray Diffraction Patterns

The actual pattern resulting from powdered samples is similar to the single crystal experiment, in which the data are generated from the particular arrangement of atoms within the unit cell.

Factors which affect the intensity and number of peaks (reflections) include:

1. Crystal class
2. Lattice type
3. Symmetry
4. Unit cell parameters
5. The distribution and type of atoms in the unit cell

As a result of the enormous range of different structures which materials

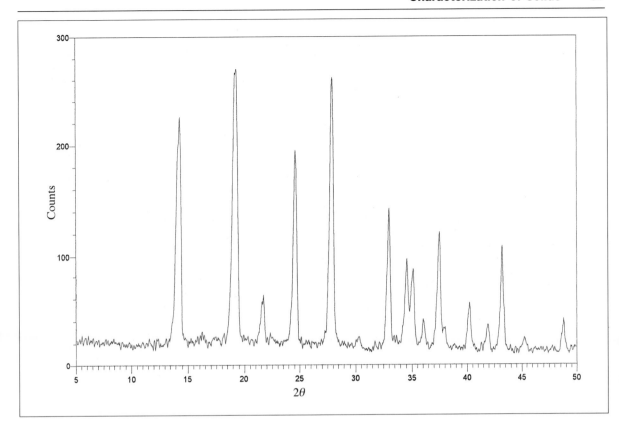

adopt, nearly all crystalline solids have a unique powder X-ray diffraction pattern in terms of the *positions* and *intensities* of the observed *reflections* (not peaks). Even in mixtures of compounds, each phase will have its own set of reflections, so the relative intensities of one set of lines compared with another depends to a certain extent on the amount of each phase present.

Figure 3.9 Powder X-ray diffraction pattern

3.1.5 Uses of Powder X-ray Diffraction Data

Identification of Unknown Materials

Over 150,000 unique powder diffraction data sets have been collected from organic, organometallic, inorganic and mineral samples. These have been compiled into a database known as the JCPDS (Joint Committee on Powder Diffraction Standards). The experimental diffraction data can be cross-matched by on-line search or by comparing the three most intense reflections. This methodology is used as a quick method for the identification of impure materials, and it is also used widely in industry, *e.g.* for quality control in drug samples or the partition of minerals in clays.

Phase Purity

As all crystalline phases will contribute to an overall powder X-ray diffraction pattern, the X-ray technique is routinely used to follow reactions and to monitor the purity of products. For example, the reaction of lanthanum oxide with iron oxide is shown in Figure 3.10. At the start of the reaction the pattern is just a combination of that of La_2O_3 and Fe_2O_3. After some time, a new set of reflections appear, corresponding to the product, $LaFeO_3$. On completion of the reaction the only set of reflections remaining is that of the product. Phase purity can be monitored in this way down to a level of *ca.* 3%.

However, there are limitations to the technique when assessing phase purity and also in general. Since X-rays are scattered by electrons, light elements with few electrons scatter poorly in comparison with heavy elements. Therefore, if several phases are present and one contains a heavy element and the others do not, *e.g.* lead carbonate mixed with magnesium oxide and carbon, then the former will produce a much

Figure 3.10 Reaction of iron oxide (bottom) with lanthanum oxide (second from bottom) to form lanthanum iron oxide (top three patterns)

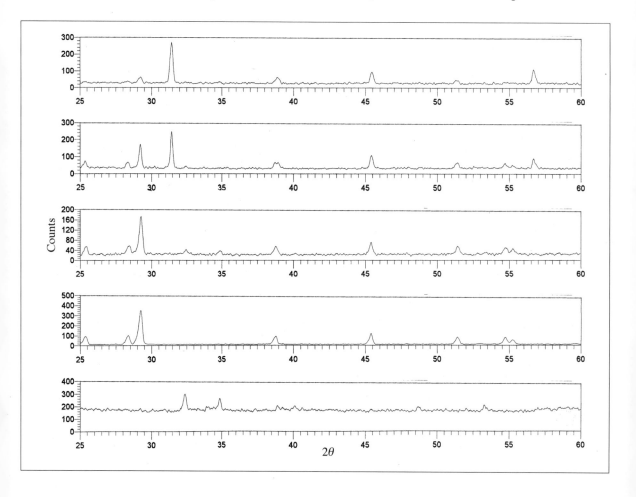

2θ

stronger pattern. Secondly, reflections giving diffraction patterns are only observed from *crystalline* phases. No matter how much *amorphous* substance is present, it will not produce a diffraction pattern except a diffuse halo in the background, as shown in Figure 3.11.

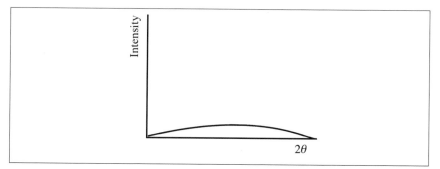

Figure 3.11 Scattering from a non-crystalline sample

Determination and Refinement of Lattice Parameters (Indexing)

If Miller indices can be assigned to the various reflections in the powder pattern, it becomes possible to determine the cell constants. This assignment is readily achieved for cubic crystal systems with a simple relationship between the diffraction angles and lattice parameters.

The equation relating the diffraction angle to the Miller indices is given below:

$$\sin^2\theta = \frac{\lambda^2}{4a^2}(h^2 + k^2 + l^2)$$

The wavelength and the cell parameter are just constants, so this can be rewritten as:

$$\sin^2\theta = C(h^2 + k^2 + l^2)$$

Dividing the $\sin^2\theta$ values of all of the other reflections by that of the first reflection removes C and gives a number corresponding to the ratio of the Miller indices. If the first reflection is at an angle 2ϕ with Miller indices h,k,l, and a general reflection is at 2θ with Miller indices $h_1,k_1,+l_1$, then:

$$\frac{\sin^2\theta}{\sin^2\phi} = \frac{h_1^2 + k_1^2 + l_1^2}{h^2 + k^2 + l^2}$$

Using this procedure for a set of reflections in an unknown cubic material gives the columns of data tabulated in Table 3.1. For a primitive lattice the first reflection is the 100; therefore the ratio is just:

$$\frac{\sin^2\theta}{\sin^2\phi} = \frac{h_1^2 + k_1^2 + l_1^2}{1^2}$$

So the values of h_1, k_1 and l_1 can be calculated.

Worked Problem

Q Given that dividing the $\sin2\theta$ value for the 48.266 reflection in Table 3.1 by the first reflection gives 6, what are h, k and l?

A
$$\frac{\sin^2\theta}{\sin^2\phi} = \frac{h_1^2 + k_1^2 + l_1^2}{1^2} = 6$$

So we are simply looking for three squared numbers which add up to six, i.e. 2, 1 and 1, because $2^2 + 1^2 + 1^2 = 6$. Thus the reflection is the 211. It is equally correct to say the 121 or the 112 as the material is cubic and a, b and c are the same, but by convention the highest number normally comes first.

Table 3.1 Indexing powder diffraction data

2θ	$\sin^2\theta$	Ratio	Miller indices
19.213	0.0279	1	100
27.302	0.0557	2	110
33.602	0.0836	3	111
38.995	0.1114	4	200
43.830	0.1393	5	210
48.266	0.1671	6	211
56.331	0.2228	8	220
60.093	0.2507	9	300
63.705	0.2785	10	310
67.213	0.3064	11	311
70.634	0.3342	12	222

Repeating this procedure for all the reflections is called indexing of the data, assigning Miller indices to the 2θ reflections.

Note that there are some values which are not possible for the sum of the indices, e.g. 7 or 15, as there is no combination of the squares of three *integral* numbers which adds up to 7 or 15.

After indexing, one of the peaks can be used to calculate the cell parameter if the wavelength is known. As the error in measuring the 2θ angle is the same for any reflection (systematic error), the last reflection is used as it has the smallest percentage error.

Worked Problem

Q Given that the X-ray wavelength was 1.54 Å, use the 222 reflection in Table 3.1 to calculate the lattice parameter, *a*.

A

$$\sin^2\theta = \frac{\lambda^2}{4a^2}(h^2 + k^2 + l^2)$$

Therefore

$$0.3342 = \frac{1.54^2}{4a^2}(2^2 + 2^2 + 2^2)$$

Giving *a* = 4.613 Å for the unit cell parameter.

Lattice Type and Systematic Absences

The pattern we have just indexed is for a primitive lattice where *all* reflections are observed. In the other types of cubic lattice, certain types of reflection are absent and these are called systematic absences.

The origin of the absences is due to destructive interference occurring between the diffracted waves, meaning that the intensity cancels out. For example, Figure 3.12 shows diffraction from the 100 plane of a body-centred cubic lattice.

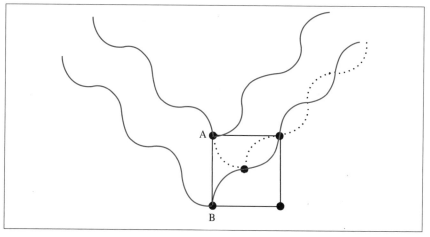

Figure 3.12 Diffraction from the 100 plane in a body-centred lattice

The scattering from A and B (the atoms at the corners of the cell) will always be in phase, but there is also an atom at the centre of the cell. Diffraction from this atom will be exactly half a wavelength out of phase with the diffracted beam from the atoms at the corners. As we know from atom counting in Chapter 1, these sets of atoms occur in pairs throughout the lattice (*i.e.* $Z = 2$). This means that there will be total destructive interference for the 100 reflection and no intensity will be seen. For any body-centred (I) lattice this condition will hold. Extension of this principle to other reflections generates the following rule. *For a reflection to be seen for a body-centred lattice, then the sum of the Miller indices must be even*:

$$h + k + l = 2n$$

So, the 110 reflection would be seen, but the 111 would be absent.

The analogous condition for a face-centred cubic lattice is:

h,k,l must be all odd or all even

So here the 110 is absent, but the 111 will be seen.

This can lead to problems in the indexing procedure, as for a body-centred lattice the first reflection is *not* the 100 but the 110.

How Do You Know When the Lattice is Not Primitive?

If the lattice you are trying to index is body centred rather than primitive, then when you take the ratios a seven will appear in the list, which is equivalent to taking the ratio of the 321 reflection over the 110 reflection. As no three squared integers add up to seven, this is the best indication that the lattice is body centred, and the ratios must be multiplied by two before the assignment of Miller indices can begin.

If the ratios appear as thirds, then the lattice is face centred and the ratios should be multiplied by three before assigning the indices (where the first reflection allowed in a face-centred cubic lattice is the 111 and the second the 200, giving 1.3333 as the ratio).

Worked Problem

Q Reaction of buckminsterfullerene with potassium affords a compound which contains 75.4% carbon and 24.6% potassium. The powder diffraction pattern recorded with copper radiation gives the following 2θ reflections: 10.97, 15.54, 19.06, 22.05, 24.69, 27.09,

29.30, 31.38. Determine the lattice type and calculate the lattice parameter.

A See Table 3.2. The seven in the ratio results from the lattice being *body centred*, where

$$\frac{\sin^2\theta}{\sin^2\phi} = \frac{h_1^2 + k_1^2 + l_1^2}{1^2 + 1^2 + 0^2} = \frac{h_1^2 + k_1^2 + l_1^2}{2}$$

Table 3.2 Ratio of $\sin^2\theta$ for K_6C_{60}

2θ	$\sin^2\theta$	Ratio	$\times 2$
10.97	0.0091	1	2
15.54	0.0182	2	4
19.06	0.0274	3	6
22.05	0.0366	4	8
24.69	0.0457	5	10
27.09	0.0549	6	12
29.30	0.0640	7	14
31.38	0.0730	8	16

To produce a ratio which is the sum of the squares of the unknown Miller indices requires the ratios to be multiplied by two. Using the largest $\sin^2\theta$ value, the lattice parameter can be calculated. Now

$$0.0730 = \frac{1.54^2}{4a^2}(16)$$

Giving $a = 11.400$ Å.

Isoelectronic Elements

The scattering of the X-ray beam is directly proportional to the number of electrons an element has. In any X-ray-based technique, elements which are isoelectronic (have the same number of electrons) are indistinguishable to X-rays. This can cause problems in assigning sites in single crystal diffraction, and can lead to additional absences in the powder X-ray diffraction pattern.

For example, MgO contains the magnesium cation and the oxide anion, which both have ten electrons. MgO forms the halite face-centred cubic lattice with alternating magnesium cations and oxide anions throughout the structure. The reflections actually observed are for a primitive lattice with a lattice parameter of half the size. This is because the unit cell observed by the X-rays views the two species as the same. Similar problems arise with many neighbouring elements such as phosphorus and chlorine, and assignments of such atoms to a particular site typically is achieved by considering the bond lengths, angles and coordination geometry.

Structure Refinement

In the powder experiment all the reflections occur along the same axis and so they frequently overlap. Extracting the intensities of individual reflections therefore becomes very difficult, so the pattern is refined as a whole rather than using the reflections individually. This is called the Rietveld method.[4–7]

A trial structure is postulated and a theoretical pattern generated from the trial structure and compared to the experimental data. By refinement of the trial structure by moving atoms, changing lattice parameters, *etc.*, the trial structure is then changed until the theoretical pattern becomes a good match to the experimental data.

Although very complex structures can be solved in this way, a trial structure is always needed to start the process. This generally means that for new structures a single crystal determination is needed.

Crystallite Size

In order to observe sharp diffraction maxima in the powder X-ray diffraction experiment, the crystallites must be sufficiently large so that slightly away from the 2θ maximum, destructive interference occurs and the intensity returns to background level. This condition is not met when the crystallites are too small, there are few diffraction planes and the reflection broadens. Figure 3.13 shows the effect of decreasing particle size on the width of the reflections.

The Scherrer formula relates the thickness of a crystallite to the width of its diffraction peaks, and is widely used to determine particle size distributions in clays and polymers:

$$t = \frac{0.9\lambda}{\sqrt{B_M^2 - B_S^2} \cos\theta}$$

Where t is the crystallite thickness (Å), B_M and B_S are the width in radians of the diffraction peaks (at half maximum height) of the test sample and a highly crystalline standard sample, respectively, and λ (Å) is the wavelength of the X-ray beam.

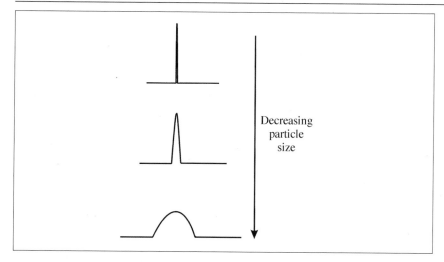

Decreasing
particle
size

Figure 3.13 Evolution of peak width with decreasing particle size

3.2 Neutron Diffraction

Neutrons in thermal equilibrium at 298 K can be used for diffraction in a similar way to X-rays, since they also have wavelengths comparable to interatomic spacings. In contrast to X-ray diffraction, the powder neutron diffraction experiment is much more common than single crystal neutron diffraction, since the beam intensity tends to be 1000 times less than for X-ray diffraction, so that single crystals of a sufficient size to collect good data are difficult to grow.

The neutron technique is very much a complementary technique to the X-ray experiment as neutrons can interact very differently with isotopes and also, unlike X-rays, interact strongly with light elements. The neutron diffraction experimental technique varies slightly and is described below, but the general principles are similar to the X-ray experiment.

3.2.1 Experimental Methods

The Bragg equation is utilized in two different ways in two sorts of powder neutron diffraction experiment:

1. Fixed wavelength (similar to X-rays), using a reactor source.
2. Fixed theta (angle of incidence), using a spallation source.

Reactor Source

High-energy neutrons are created by the nuclear fission process from a uranium target:

$$U + n \rightarrow Fragments + 3 \; neutrons$$

This produces *high-energy* neutrons of *variable wavelength,* but a single wavelength of much lower energy is needed for the diffraction experiment. Firstly, the overall energy is reduced by passing the beam into a moderator (which is a fluid or solid kept at constant temperature); here the neutrons undergo inelastic collisions, resulting in the loss of energy until thermal equilibrium is attained. The energies of the longer wavelength neutrons now follow a Boltzmann distribution, the maximum of which is dependent on the moderating fluid. To perform an experiment analogous to X-ray diffraction, a single wavelength must be selected from the band available. According to Bragg's law, if a beam of neutrons is reflected off a *single plane* (fixed *d*) of atoms at a *fixed angle*, then a single wavelength will be transmitted:

$$\lambda = 2d \sin\theta$$

This is achieved in practice using a *double single-crystal monochromator* (Figure 3.14) which uses two single crystals, parallel to each other, at a fixed angle to transmit a single wavelength. The now monochromatic beam can be used for diffraction and is scattered in a way similar to a powder X-ray experiment, with intensity as a function of scattering angle (2θ) as in Figure 3.9.

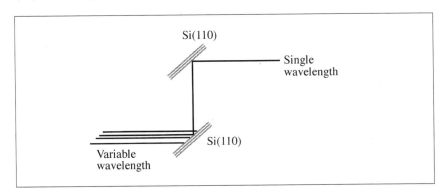

Figure 3.14 Double single-crystal monochromator

Spallation Source

The reactor source experiment is extremely wasteful as most of the neutrons are discarded at the monochromation stage. Spallation sources use all the neutrons produced after the moderation stage, but the neutrons are first produced in a slightly different way.

Hydrogen gas is fed into a ring and ionized to H^-. The H^- ions are then accelerated around a ring and the electrons stripped off to give a high-energy proton beam. The proton beam is then pulsed at 50 Hz using an alumina valve towards a uranium (or rarely tantalum) target. In the same way as in the reactor-based experiment, fission products and variable wavelength neutrons are produced from the uranium target, which are then moderated. The entire moderated beam is then used and the diffraction pattern recorded as a function of the time of flight, as seen in Figure 3.15.

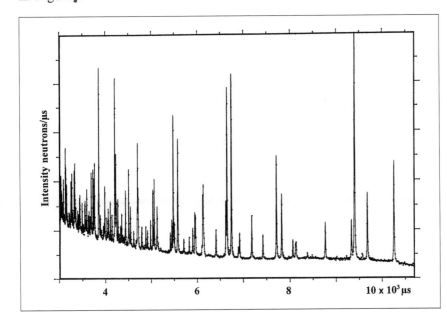

Figure 3.15 Time-of-flight powder neutron diffraction pattern

As distance (D) = time (t) × velocity (v), and from the De Broglie relation $\lambda = h/mv$, then:

$$\lambda = \frac{ht}{Dm}$$

Thus

$$\lambda = (\text{constant}) \times t$$

3.2.2 The Differences between the Behaviour of X-rays and Neutrons

Scattering Power

Variation with angle. The source of X-ray scattering is via interaction of the X-ray photon with the electron cloud of the scattering atom by electrostatic force over a distance of 10^{-10} m. Unfortunately, this distance is comparable both to the X-ray wavelength and the dimensions of the electron cloud, such that scattering from different parts of the cloud is not always in phase. This gives rise to the form factor, which is a tail-off in intensity as the size of the scattering angle increases (Figure 3.16).

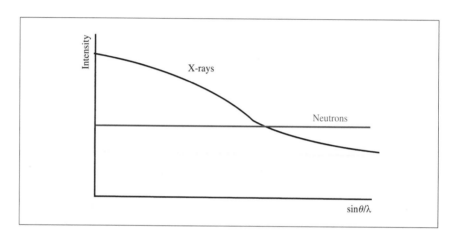

Figure 3.16 Variation of scattering amplitude with angle

Neutrons interact with the *nucleus* and because they possess zero charge the interaction occurs over a much shorter distance (10^{-13} instead of 10^{-10} m). The nucleus acts as a point scatterer, and there is no form factor as a function of angle. The difference in scattering as a function of angle is shown in Figure 3.16.

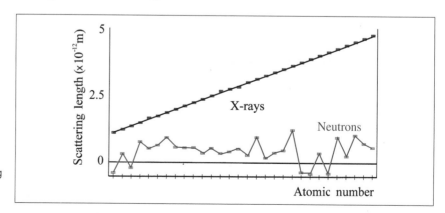

Figure 3.17 Variation of scattering length for X-rays and neutrons with atomic number

Variation with element. X-rays are scattered by electrons; therefore the more electrons something possesses, the better scatterer it will be. To a first approximation, scattering is proportional to the atomic number (Z). Neutrons are scattered by the nucleus, and this is dependent to a first degree on the size of the nucleus (*i.e.* the cross section it presents to the incoming neutron). Nuclei vary very little in size throughout the Periodic Table, so to a first approximation neutron scattering is nominally proportional to $Z^{1/3}$. However, this is not the full neutron story; superimposed on the scattering amplitude as a function of the atomic number is the resonance scattering. In this the neutron is absorbed by the nucleus and released later. This is a random effect and varies from isotope to isotope and can even lead to negative scattering, as shown in Figure 3.17.

3.2.3 Neutron Experiments

Neutron diffraction allows the following three types of experiment, which are impossible (or nearly impossible) with X-ray diffraction.

Distinguish Isotopes and Neighbours

Resonance scattering means that neighbours and isotopes can have very different scattering lengths, *e.g.* two nickel isotopes, ^{55}Ni (0.87) and ^{62}Ni (1.44), or the transition elements Fe (0.95), Co (0.25) and Ni (1.02) (all $\times 10^{-12}$ m), whereas for X-rays, scattering would be closely proportional to the atomic numbers 26, 27 and 28, respectively. Neutron diffraction is therefore particularly useful in the study of alloys.

Similar Scattering for All the Elements

Relative scattering in X-rays and neutrons is given in Table 3.3 for some very different elements. In neutron diffraction, all elements have a scattering amplitude of the same order of magnitude, *i.e.* between -0.4 and 1.6×10^{-12} m. The similar scattering lengths for neutron diffraction for all the elements means that the scattering of light elements (*e.g.* D) can be detected in the presence of heavy ones (*e.g.* U).

Real Bond Lengths

X-rays are scattered by the electron cloud, so polarized bonds distort the cloud and a true representation of the atomic position is not given. The electron density is distorted towards the electronegative element, and the measured distance is always too short. For example, O–H typically is measured at 0.8 Å, which is less than the sum of the ionic radii! Neutron

Table 3.3 Comparison of relative X-ray and neutron scattering[8]

Element	X-ray	Neutron
Deuterium	1	0.667
Phosphorus	15	0.513
Sulfur	16	0.285
Chlorine	17	0.958
Tin	50	0.623
Tungsten	74	0.486

diffraction measures the distance between the nuclei, and so the true value of *ca.* 0.96 Å is measured.

3.2.4 Bond Valence Calculations[1,9]

The method of assignment of single, double and triple bonds which is applicable to organic chemistry cannot be used in crystalline, non-molecular chemistry. To overcome this problem, an empirical method has been developed by a number of workers to describe bonds in non-molecular solids. In a similar way to valence bond theory, this model attempts to assign valences or strengths using the individual bonds. The difference between the two methods is that this procedure allows non-integral values to be assigned, where the valence is summed by consideration of all the bonds around a particular atom, using the equation:

$$s_{ij} = \exp\left(\frac{r_0 - r_{ij}}{B}\right)$$

where s_{ij} is the valence of a particular bond, r_{ij} is the measured bond length and r_0 is the bond length unit derived for unit bond valence (*i.e.* a single bond between atoms i and j); B is a constant (0.37). Determining the valence is then just a matter of summing over all the bonds. The calculated effective valence should be within 0.2 of the expected valence for the atom. For example, for Sr, which normally exists in the 2+ oxidation state, the calculated bond valence should lie between 1.8 and 2.2. This procedure provides a method which allows new structures to be checked for chemical sense, where all atoms in the crystal structure should obey the valence sum rule.

3.3 Electron Microscopy

The electron microscope has become a very important part of materials structure determination, yielding information on both the morphology, surface structure and element composition of new or known phases. The very small amount of material needed for such experiments means that electron microscopy can be used not only for fault analysis and routine determinations on bulk materials, but also on single crystals.

The principle involves using X-rays (beam of electrons) to image atomic structures. The very short wavelengths allow much higher resolution than with a standard optical microscope. There are several types of electron microscope which allow different types of image to be formed. They operate in either transmission mode (electrons pass through the sample) or reflection mode (electrons are reflected from the surface)

3.3.1 Transmission Electron Microscope

Transmission electron microscopy (TEM) uses both elastic and inelastic scattering of the electrons to form an image which can be interpreted either on screen or on a photographic plate. The image produced has no depth profile, as the electrons pass through the sample; so surface features cannot be investigated, and the area viewed is fixed but is well resolved. The image available from TEM is very dependent on how the sample is prepared, as very thin sample films are required (~2000 Å) to produce a good image. These films can be achieved in a number of ways, including ion bombardment. TEM allows crystal defects such as dislocations, stacking faults and phase boundaries to be seen directly. In addition, with appropriate stages the effect of high and low temperature or chemical treatment on the sample can be observed *in situ*. The TEM readily resolves images in the range 10–1000 Å.

3.3.2 Scanning Electron Microscope

Scanning electron microscopy (SEM) uses the electrons reflected from the surface of a material to form the image, such that the thickness of the sample is not important and sample preparation is less difficult than for TEM. The probe produces a high-intensity focused electron beam, which backscatters from the surface and is continuously measured as the microscope moves from point to point over the surface (similar to a television screen). In a non-conducting sample the surface is coated with gold or graphite to prevent build-up of surface charge. SEM can be used to image over a very large magnification range from 1 μm (100 Å) to 100 μm, allowing study of particle morphology and size, surface texture and detail and defects in surfaces such as faults and cavities caused by, for example, etching or corrosion.

3.3.3 Scanning Transmission Electron Microscopy (STEM)

This technique combines the features of the TEM (high resolution) and the SEM (surface scanning) to produce a superior instrument. The STEM can be used not only for imaging but also for analytical work; since several scanning signals are collected simultaneously, the contrast is enhanced relative to the standard instruments. The high-resolution picture gives information on structure, absorbed species and depth.

The X-rays emitted from the surface of a sample in both SEM and TEM can be used diagnostically to determine the partition of species, since they are characteristic of the elements in the sample. This process is known as EDAX (energy dispersive analysis of X-rays) and can be performed not only on areas of sample but also on individual crystallites, although light elements (with atomic numbers less than sodium) cannot be assessed as the X-rays produced are too soft and are easily absorbed.

3.4 X-ray Absorption Spectroscopy

When a beam of X-rays travels through matter (crystalline or non-crystalline) it loses intensity through interaction with the material. A proportion of the beam is absorbed as a function of increasing energy, but on top of this smooth variation is a series of sharp jumps which are caused by ejection of a core electron. These steps are characteristic of the both the element and the oxidation state of that element, and have been used for studying materials for a long time. However, superimposed on these absorption edges, from just before the edge to some 30 eV beyond it, are sharp spikes and oscillations which also contain a lot of information about the material under study. These are caused by electronic transitions between bound states (before the edge) and backscattering effects (beyond the edge), as shown in Figure 3.18.

3.4.1 X-ray Absorption Near-edge Structure (XANES)

The computational analysis of the XANES region is very complex, and hence it is often used mainly for qualitative comparison to give information on both oxidation states and coordination numbers. It is particularly used in the analysis of transition metal oxidation states and coordination numbers in catalysts

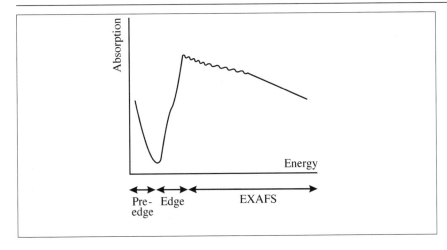

Figure 3.18 Variation of absorption with incident X-ray photon energy

3.4.2 Extended X-ray Absorption Fine Structure (EXAFS)[10,11]

Once an electron has been displaced by an X-ray photon, the ejected electron can be considered as an outgoing spherical photoelectron wave which can be backscattered by any neighbouring atoms. Superimposition of the outgoing and backscattered waves will give rise to interference effects in the final state which can be constructive or destructive (Figure 3.19). As the X-ray energy changes, so does the wavelength of the photoelectron wave, causing an oscillation between constructive and destructive interference. These oscillations can be analysed to give information on the number of backscattering atoms and the distance between them.

Figure 3.19 Constructive and destructive interference

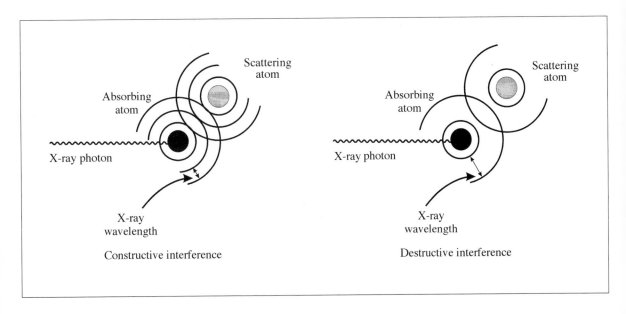

These effects occur irrespective of phase (gas, solid, liquid), temperature and crystalline state, and occur on a very short timescale; so gases, working catalysts, amorphous materials and disordered solids can be studied.

3.5   Solid State (Magic Angle Spinning) Nuclear Magnetic Resonance[12]

Solution phase NMR is a common technique applied to many molecular compounds on a routine basis.[13,14] Although similar experiments can be performed on solid samples, the bands tend to be very broad (Figure 3.20) as a result of incomplete averaging of the external magnetic field owing to the shielding created by the other nuclei in the sample (this effect is termed chemical shift anisotropy).

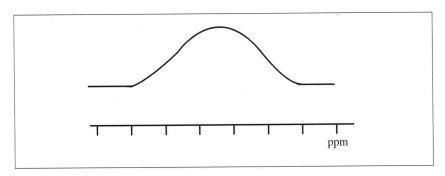

Figure 3.20 Static solid state NMR spectrum

In order to recreate the molecular tumbling which occurs in solution and which creates an averaged magnetic field, the solid is spun at the so-called **magic angle** of 54° 44′ and is derived from the equation:

$$3 \cos^2\theta - 1 = 0$$

The spinning eliminates the chemical shift anisotropy provided the spinning frequency is close to the frequency spread of the signal. Typically, a spinning rate of between 2 and 25 kHz is used. When the spectra are particularly broad and the spinning rate cannot match the spread of the signal, a series of spinning side bands are observed (such as for aluminium or vanadium), which are separated by the spinning frequency (Figure 3.21). In order to be sure which resonance is the central band and which are the spinning side bands, the spectrum is then collected at two speeds where the spinning side bands move with changing speed but the central resonance remains fixed.

In general, MAS (magic angle spinning) NMR spectra also take longer to record, as spin relaxation is slow. This is particularly true for spin-$\frac{1}{2}$ nuclei such as silicon, where relaxation times of between 120 to 600 s

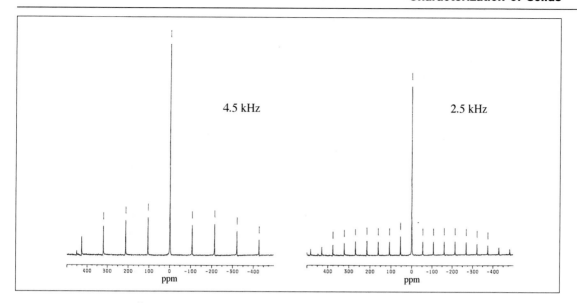

are not unusual between pulses. For quadrupolar nuclei, with more relaxation pathways, pulsing can be faster.

When the nucleus under study has poor abundance, such as for carbon-13, the cross polarization technique is used. This uses a complex pulse sequence to transfer the polarization from a nucleus with high abundance (normally hydrogen, but phosphorus and aluminium are also used) to the second nucleus, enhancing its signal.

MAS NMR is particularly suited to the characterization of zeolites,[15] where the similarity of aluminium and silicon means that the structural make-up cannot be determined by any other method. Here the position of the silicon resonance is crucial in determining the environment, with a move to more positive chemical shift as the Si–O–Si links are replaced by Si–O–Al links (referred to as Q0 through to Q4, for zero through to four aluminium atoms in the immediate shell). For example, the spectrum for zeolite P is shown in Figure 3.22.

Figure 3.21 ^{9}Be MAS NMR run at 2.5 and 4.5 kHz

3.6 Thermal Analysis

Thermal methods are used in many areas of chemistry to determine both phase changes as a function of temperature and to determine unknown quantities, such as levels of hydration or oxygen content. Although thermal analysis could be thought of as any technique which measures a property of material as a function of temperature, only the two major thermal analysis techniques will be discussed.

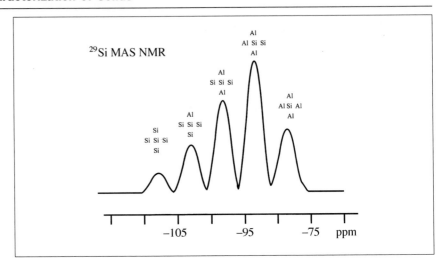

Figure 3.22 Solid state NMR spectrum of zeolite P

3.6.1 Thermogravimetric Analysis (TGA)

This technique monitors the change in weight of a sample, as it is heated in gas/air, as a function of temperature. Providing the final products are known, the difference between the final and the starting compositions can be determined, where the additional atoms have been evolved as gas/water.

Worked Problem

Q The TGA trace for the decomposition of $Al_2Si_2O_7.xH_2O$ in air is given in Figure 3.23. Given that after heating the final product is $Al_2Si_2O_7$, determine the value of x.

A

$$\frac{26.4}{RMM\ Al_2Si_2O_7.xH_2O} = \frac{22.7}{RMM\ Al_2Si_2O_7}$$

If the relative formula mass of $Al_2Si_2O_7$ is A, then rearranging gives:

$$x = \frac{A(26.4 - 22.7)}{(22.7)(18)}$$

Therefore $x = 2$.

3.6.2 Differential Thermal Analysis (DTA)

When a material goes through a phase change during the heating cycle, the act of changing phase costs energy, such that energy is put in but the temperature of the sample remains constant. If a sample is heated against a standard sample, which exhibits no such change, then a differential between the two is apparent whilst the active sample melts or changes phase. A DTA curve for kaolinite decomposition (trough, endothermic) and recrystallization to mullite (peak, exothermic) is shown in Figure 3.23.

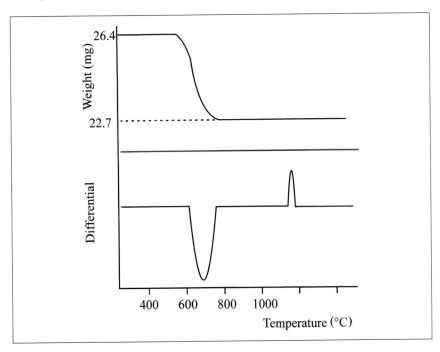

Figure 3.23 TGA and DTA spectra of the decomposition of kaolinite as a function of temperature

3.7   Other Spectroscopic Techniques

Clearly, both vibrational and UV–visible spectroscopy also play an important role in materials chemistry, and are well described in other texts.[16,17] The principles involved in carrying out these experiments on solids rather than in solution are similar, but often experimental methods vary. For example, an IR spectrum of a zeolite would be carried out by dispersing the solid in a matrix of potassium bromide and pressing into a disk, rather than in solution. Typically, a UV–visible spectrum of a solid would be carried out in diffuse reflectance mode, where the solid is dispersed in a white matrix (such as barium carbonate) and the UV light is reflected off the surface rather than passing through a solution.

These methods give the same characteristic lines you would expect from a solution species, for example a hydroxide band would be expected at 3300–3600 cm^{-1}. However, the bands are often much broader for solids.

Summary of Key Points

1. *Powder diffraction methods* are the crucial techniques involved in materials chemistry. They are used in the determination of structure, phase purity and in the monitoring of reactions.

2. *Powder diffraction patterns* can be indexed to determine cell constants in simple crystal systems.

3. *Neutron diffraction and X-ray diffraction* are complementary, *e.g.* powder X-ray diffraction to find heavy atom positions, and powder neutron diffraction to find light atoms.

4. *Electron microscopy* can provide useful information on structure, morphology and chemical composition.

5. *Spectroscopic methods* such as IR/Raman are used routinely in the characterization of solids, but tend to be used to complement other techniques where bands are often too broad to be used diagnostically.

Problems

1. Draw the 101 plane of an orthorhombic unit cell and mark on the distance d_{101}.

2. Given that the lattice parameter of primitive cubic zeolite is 12.400 Å, calculate the 2θ positions of the 301, 400 and the 111 reflections determined by copper radiation ($\lambda = 1.54$ Å).

3. β-Tungsten crystallizes with a body-centred cubic structure. What h,k,l values would you expect the first ten reflections to have?

4. The spinel $ZnFe_2O_4$ has reflections at the following 2θ values in its neutron diffraction pattern: 22.593, 26.145, 37.311, 44.059, 46.129, 53.792. Determine the lattice type, and given that the neutron wavelength was 1.90 Å, calculate the lattice parameter.

5. (i) Potassium chloride crystallizes with a face-centred cubic lattice similar to sodium chloride. If the X-ray powder pattern was recorded, what restrictions, if any, would be expected on h,k,l?
(ii) If the powder neutron diffraction pattern was recorded instead, would the restrictions change?

6. On a TGA, 24.20 mg of $SrFeO_{3-x}$ decomposed in a stream of hydrogen at 700 °C to give 23.34 mg of a mixture of Fe_2O_3 and SrO. Use these data to determine the value of x.

7. What type of cubic lattice would have a diffraction pattern where the 110 reflection would be observed but the 234 would be absent? Give reasons for your answer.

8. $LiSrH_3$ crystallizes as a white cubic solid after the reaction of lithium and strontium under a flow of hydrogen at 500 °C.
(i) What factors affect the scattering of X-rays by crystalline solids? Why might this compound be difficult to characterize?
(ii) The powder X-ray diffraction pattern of this materials shows the following reflections:

$$2\theta = 23.25, 40.84, 47.53, 53.56, 69.49, 74.38$$

Index the data, determine the lattice type and calculate a cell parameter ($\lambda = 1.54$ Å.).

References

1. I. D Brown and D Aldermatt, *Acta Crystallogr., Sect. B*, 1976, **32**, 1957.
2. W. Clegg, *Crystal Structure Determination,* Oxford University Press, Oxford, 1998.
3. M. F. C. Ladd and R. A. Palmer, *Structure Determination by X-ray Crystallography,* 3rd edn., Plenum, New York, 1993.
4. H. M. Rietveld, *J. Appl. Crystallogr.,* 1969, **2**, 65.
5. H. M. Rietveld, *Acta Crystallogr.,* 1967, **22** 151.
6. R. A. Young, *The Rietveld Method,* Oxford University Press, Oxford, 1997.
7. A. C. Larson and R. B. von Dreele, *GSAS: Generalized Structural Analysis System,* Los Alamos, NM, 1990.

8. *Neutron News,* 1992, **3**(3), 29.
9. I. D. Brown, *Chem. Soc. Rev.*, 1975, **4**, 359.
10. B. K. Teo, *Acc. Chem. Res.*, 1980, **13**, 412.
11. B. K. Teo and P.A. Lee, *J. Am. Chem. Soc.*, 1979, **101**, 2815.
12. C. A. Fyfe, *Solid State NMR for Chemists,* CRC Press, Boca Raton, FL, 1983.
13. R. K. Harris and B. E. Mann, *NMR and the Periodic Table,* Academic Press, London, 1978.
14. P. J. Hore, *Nuclear Magnetic Resonance,* Oxford University Press, Oxford, 1995.
15. G. Englehardt and D. Michel, *High Resolution NMR of Silicates and Zeolites,* Wiley, New York, 1987.
16. A. K. Brisdon, *Inorganic Spectroscopic Methods,* Oxford University Press, Oxford, 1998.
17. E. A. V Ebsworth, D. W. H. Rankin and S. Craddock, *Structural Methods in Inorganic Chemistry*, Blackwell, Oxford, 1986.

4
Preparation of Materials in the Solid State

The preparation of materials in the solid state is rather different from the synthesis of discrete molecules. The properties of inorganic and organic molecular systems can be altered by reaction in solution, to add or remove particular groups, and the products can be purified by recrystallization. The synthesis and alteration of solids is very different. It involves treatment of the whole lattice. Often, post-synthesis purification of the materials is not possible owing to the low solubility of the phases formed. Hence, every effort must be made to avoid excesses of reagents.

Aims

This chapter aims to outline particular techniques which are available for the creation of solid state materials. By the end of this chapter you should be able to:

- Identify suitable methods to prepare different solid materials
- Choose techniques to alter the 'as prepared' stoichiometry of a material
- Understand the particular methodology for preparation of single crystals from solid materials
- Distinguish between different sorts of precursor method

4.1 High-temperature (Ceramic) Method

The most widely used method for the preparation of solid materials is by *reaction of solid components in the correct molar proportions at elevated temperature over a long period*. Many mixed metal oxides, chalcogenides (sulfides, selenides and tellurides) and pnictides (phosphides and nitrides) have been prepared using this method. Accurate weighing of the starting materials is of utmost importance because the material cannot be purified after the synthesis. For this reason, the choice of reactants is important, and care must be taken not to use hygroscopic (water absorbing) or non-stoichiometric (variable composition) starting materials.

Worked Problem

Q The target oxide is $BaNiO_2$; what reactants would be a suitable choice?

A Many starting materials could be used, but barium oxide (BaO) is not a suitable choice. BaO is highly hygroscopic and absorbs carbon dioxide from the air. This means an amount of $BaO/Ba(OH)_2/BaCO_3$ would be weighed out rather than the molar quantity of BaO required. The mixture of starting materials would be deficient in barium, and hence, so would the product. In cases like these, where the logical choice is unsuitable, a metal carbonate or nitrate would be used. These metal salts decompose stoichiometrically to the oxide on heating:

$$BaCO_{3(g)} + NiO_{(s)} \rightarrow BaNiO_{2(s)} + CO_{2(g)}$$

Care should also be taken to avoid volatile or air-sensitive compounds, if possible. In cases where there is no alternative, **sealed tube methods** (Section 4.3) can be used.

Worked Problem

Q A series of alkali metal titanates are to be prepared. What preparative conditions could be suitable?

A Solid state reactions occur at high temperature, and materials which are often thought of as non-volatile can become volatile under these conditions. For example, the heavier alkali metals such as caesium are volatile above 600–700 °C. To overcome this problem, reactions using them are either performed in a sealed tube, to avoid loss of the reactants, or with at least a 10% excess of the alkali metal salt.

Since solid state reactions can require temperatures up to 2000 °C, conventional glassware is often not suitable. For example, borosilicate glass (Pyrex) softens above approximately 600 °C. Instead, high melting point materials such as precious metal capsules (gold, platinum), alumina and zirconia crucibles (Figure 4.1) and silica are used as containers in solid state synthesis. Vessel selection depends on the reactants and the temperature, as shown in Table 4.1. In some cases, vessels are coated with graphite before use to prevent reaction with the material under study, as at such high temperatures even the containers can become reac-

Figure 4.1 Alumina and zirconia crucibles
(Kindly donated by John Savage, Technical Glass Company)

Table 4.1 Containers for materials samples

Vessel	Maximum temperature (°C)*	Melting point (°C)
Borosilicate glass (Pyrex)	515	820
Gold tubing	1013	1063
Silica (quartz)	1405	1853
Platinum	1719	1769
Alumina (Al_2O_3)	1900	2072
Zirconia (ZrO_2)	2000	2700
Magnesia (MgO)	2400	2852

* Nominal working temperature for empty container; actual working temperature will be sample dependent.

tive. This is achieved by using a small amount of a low boiling point solvent such as propanone, surface wetting, and then heating to white heat using a blow torch to decompose the organic solvent to carbon. This method is particularly effective in preventing the reaction of alkali metals and alkaline earth metals with glass.

The usual laboratory methods of heating, such as using mantles or ovens, do not provide high enough temperatures to allow the solids to react. Very high temperatures (4000 °C) can be achieved using lasers, but most reactions do not require such extreme conditions. Generally, temperatures up to 2000 °C are achieved in furnaces which use resistance heating to reach the required temperature. Examples of the two sorts of furnace are shown in Figure 4.2. The box (or muffle) furnace heats a large volume to a fixed temperature. The tube furnace has a smaller heated area at the centre of the tube (typically 2–5 cm) and the temperature falls off rapidly towards the ends of the tube. The tube furnace can be used to heat samples under different gases by placing inside it a tube connected to a gas cylinder. Single crystals can also be grown in tube furnaces (Section 4.6).

The majority of useful complex materials are oxides, but the ceramic method can also be used to prepare metal halides and nitrides by using controlled atmospheres and appropriate reaction vessels. For example, the formation of calcium gold nitride from calcium nitride and gold[1] in a sealed gold capsule under nitrogen:

$$2Ca_3N_{2(s)} + 3Au_{(s)} \rightarrow 3Ca_2AuN_{(s)} + \tfrac{1}{2}N_{2(g)}$$

4.1.1 Reactions in the Solid State

Ceramic methods are very slow. This is mainly because, with no melt formed, the entire reaction occurs in the solid state and requires diffusion across the points of contact in a mixture. For example, Figure 4.3 shows a schematic diagram of the reaction between solid particles. The reaction mixture consists of a collection of particles and voids. The cations migrate across the points of contact and form the new structure at the phase boundary (normally the cations are the mobile ions as they are much smaller than the anions, e.g. 0.95 Å for Na^+ compared with 1.4 Å for O^{2-}). As more of the new compound forms, the cations have to migrate further, through the product, to form the new phase. Therefore, the reaction becomes slower and slower, as the diffusion path gets longer and longer.

Decreasing the length of the diffusion path and increasing the number of points of contact between the particles in the mixture can speed up the reaction. This can be achieved by:

Figure 4.2 Box and tube furnaces
(Kindly donated by Pyrotherm Furnaces Ltd.)

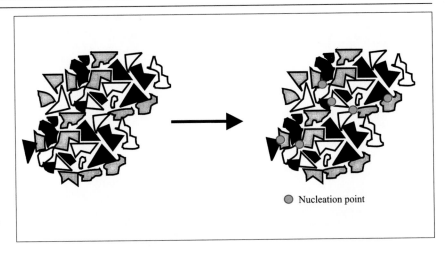

Nucleation point

Figure 4.3 Schematic diagram of the reaction of solid particles

- Regrinding the mixture periodically to improve homogeneity, after every 16–24 hours
- Improve contact between the particles by forcing the particles closer together, *e.g.* by pressing the mixture into pellets to reduce the number of voids
- Increasing the temperature to give the particles more energy to move

In practice, a combination of all three methods is used. Pressing the mixture into pellets is particularly useful, as it can also reduce the volatility of the reactants. A typical reaction scheme for the formation of $LaFeO_3$ is shown in Figure 4.4. The process of regrinding and heating is often referred to as the 'shake and bake' method. It is a feature of most reactions carried out in the solid state.

The long reaction time is only one of the problems associated with the ceramic method. Finding the right annealing conditions and temperature is primarily a matter of trial and error, as the method of preparation of *new* materials cannot be easily predicted. Often, the product is not totally homogeneous in composition, even when the reaction proceeds almost to completion. If there is any impurity left at the end of the reaction, it is normally impossible to remove.

The other way to overcome the very slow diffusion process is to decrease the particle size so the cations have a shorter distance to travel. By mechanical grinding the smallest particle size achievable is between 10,000 and 100,000 Å, which is roughly 1000–10,000 unit cells. This is a considerable distance for the diffusing cation. In the following methods, a precursor is created, normally by solvation in a suitable solvent. This allows mixing on an *atomic scale* before the solvent is removed, leaving the solids thoroughly mixed, and a much smaller particle size. The residue is then heated using the normal ceramic methodology, but the reaction temperature can often be significantly lowered.

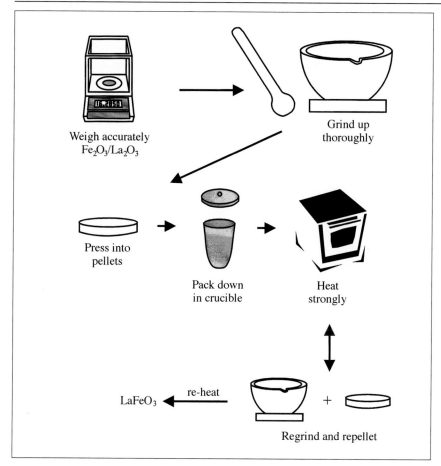

Weigh accurately
Fe_2O_3/La_2O_3

Grind up
thoroughly

Press into
pellets

Pack down
in crucible

Heat
strongly

$LaFeO_3$ ← re-heat ← Regrind and repellet +

Figure 4.4 Reaction scheme for the formation of lanthanum iron oxide using the ceramic method.

Spray and Freeze Drying

In both cases, the reactants are dissolved in a common solvent and then either sprayed into a hot chamber, where the solvent evaporates instantaneously (spray drying), or frozen in liquid nitrogen and the solvent removed at low pressure (freeze drying).

Solvation of Metal Salts

This method involves dissolving metal salts (often nitrates, hydroxides or oxalates) in a suitable solvent and then evaporating the mixture to dryness. The dried residue is then reacted as in the ceramic method. This procedure only works well if the components have similar solubilities, otherwise the solid with the lowest solubility starts precipitating out first, giving no better mixing than the normal ceramic method.

> ## Worked Problem
>
> **Q** What would be a good choice of solvent for preparing an intimate mixture of cerium oxalate and praesodymium oxalate?
>
> **A** Often a good choice of solvent for the dissolution reactions is the acid of the salt in question. In this case, oxalic acid would be a good choice.

Co-precipitation

Mixed ion solutions can be precipitated to produce a solid containing the required ions, although care must be taken to ensure that the *correct ratio* of ions precipitates. For example, oxalic acid can be used to precipitate zinc iron oxalate from a solution of the ions dissolved in acetic acid:

$$Zn^{2+}_{(aq)} + 2Fe^{2+}_{(aq)} + 3C_2O^{2-}_{4(aq)} \rightarrow ZnFe_2(C_2O_4).6H_2O$$

The spinel $ZnFe_2O_4$ is then produced by heating the precipitated mixture to 800 °C. Although this method does allow a significantly lower firing temperature to be used than in the direct ceramic method, it is particularly difficult to use for many ion systems.

Sol–Gel Methods

This method is particularly important, and is used widely in the preparation of zeolites, thin films and abrasive coatings. The process involves preparation of a concentrated solution (sol) of the reaction components. This mixture is then converted to an almost rigid gel by removing the solvent or by adding a component which causes the gel to solidify. The gel can then be treated in some way to produce the desired material. This methodology can also be used to produce composite materials, where organic and inorganic species can be trapped in an inert matrix. The properties of both the inert matrix and the trapped species are combined in the product.

4.2   Sealed Tubes

In some cases, direct reaction under ambient conditions (in air at one atmosphere pressure) to form materials cannot be performed. This may be due to a number of reasons, including volatility of the reactants, air sensitivity of the starting materials and/or products, or the desire to form a compound with a metal in an unusually low oxidation state.

For example, powdered manganese and sulfur will react together at 500 °C to form MnS:

$$Mn_{(s)} + S_{(s)} \rightarrow MnS_{(s)}$$

but if heated at this temperature in air, all the sulfur would volatilize by 110 °C. In order to keep manganese and sulfur in the same vicinity, they must be reacted together in a sealed vessel.

Typically, for this type of reaction the components are loaded into a glass or quartz ampoule in a glass box, evacuated and then sealed off by melting the glass/quartz using a blow torch. The whole tube is then heated subsequently in a tube or box furnace. The tube is then scored and broken open to remove the product after cooling. Similarly, reactions can take place in precious metal capsules sealed by welding. These reactions can be a bit hazardous as the tubes can react with the materials placed in them, which causes the tube to weaken. Significant pressure can also build up inside the tubes, which can cause minor explosions. Typically, these reactions are performed inside a work tube made of metal to prevent damage to the furnace if the reaction vessel breaks.

4.3 Controlled Atmospheres

Some materials cannot be generated by heating in air at one atmosphere pressure. To prepare such materials a controlled atmosphere is used, where a particular gas is passed over the reaction mixture in a tube furnace. For example, the only readily available oxide of vanadium is V_2O_5. To prepare VO, where vanadium is in the +2 oxidation state, requires hydrogen to be passed over V_2O_5 at approximately 1000 °C:

$$V_2O_{5(s)} + 3H_{2(g)} \rightarrow 2VO_{(s)} + 3H_2O_{(g)}$$

Alternatively, to prepare a compound containing nickel in the +3 oxidation state, e.g. $LaNiO_3$, requires oxygen annealing, since under normal conditions nickel would be in the +2 state:

$$LaNiO_{2.5(s)} \rightarrow LaNiO_{3(s)}$$

This reaction is quite characteristic of compounds with extended lattices, rather than discrete structures, in that $LaNiO_{2.5}$ is non-stoichiometric. This means that not all the available sites in the lattice are filled, giving a non-integral number of atoms in the chemical formula. Different types of non-stoichiometry, and the effect it has, are discussed further in

Chapter 6. It should be noted that non-stoichiometry in solid state compounds can affect both the preparative conditions and the physical properties of the material.

Often annealing (heat treating of a prepared phase) reactions take place at much lower temperature than the initial preparation. This is simply because samples prepared at high temperature and quenched (rapidly cooled) are often disordered, particularly when the phase is non-stoichiometric. Annealing is used for all sorts of materials to remove defects, improve homogeneity and reduce strain. For example, newly blown glassware is annealed to remove imperfections to improve strength. While phase preparation is typically carried out in a box furnace, annealing reactions are often carried out in a tube furnace equipped with a tube of ceramic or silica to allow a gas to be passed over the mixture. To create a positive pressure inside the tube, a bubbler containing a viscous liquid is sometimes placed after the work tube, as shown in the schematic drawing (Figure 4.5)

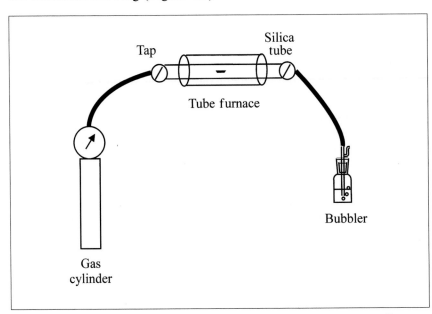

Figure 4.5 Tube furnace annealing reaction

Unfortunately, the conditions provided by tube furnace annealing are sometimes insufficient to produce high or unusual oxidation states. Higher pressures are required. For example, high-pressure nitrogen has been used to generate ternary nitrides, and high-pressure oxygen has allowed unusual oxidation states to be accessed:[2]

$$Sr_3Fe_2O_6 + O_2 \rightarrow Sr_3Fe_2O_7$$

i.e.

$$Fe^{3+} \rightarrow Fe^{4+}$$

The stored energy of high-pressure gas systems makes such reactions quite hazardous, and so they are performed on a small scale (*ca.* 1 g). The maximum pressure available for each gas is dependent on the gas in question, as phases liquefy at different pressures. The upper limit of these reactions is 18 kbar (1.8 GPa) for argon-based vessels. Here it is the absolute pressure that is the important factor, and not the *type* of gas.

4.4   Hydrothermal Methods

Relatively low-temperature solutions can be used to produce insoluble materials by direct reaction. For example, metal halides at room temperature, *e.g.*

$$3KF_{(aq)} + NiCl_{2(aq)} \rightarrow KNiF_{3(s)} + 2KCl_{(aq)}$$

Such reactions can take place at low temperature in an open vessel. However, the range of these reactions can be extended by using closed vessels and heating the solvent above its boiling point in a hydrothermal reaction. This type of reaction is particularly important in the preparation of a family of aluminosilicate materials known as the zeolites (Chapter 7). The zeolites are framework structures constructed of vertex-linked aluminium and silicon tetrahedra. They are formed from an aluminium and silicon source heated in the presence of a templating agent. The templating agent is a anion or cation of particular *shape*. The tetrahedra in solution use the templating agent to form the framework by linking together around it. For an aluminosilicate framework, the aluminium and silicon sources are typically mixed with the template and allowed to 'age' by standing the reaction mixture for a period of time before heating. During this time the molecules organize themselves into amorphous conglomerations. The pre-organized mixture then crystallizes within the conglomeration by using up the amorphous material.[3]

Worked Problem

Q What is the chemical formula of a tetrahedral unit, AX_n, which shares all vertices?

A If the unit was isolated the formula would be AX_4, but if each

X is shared between two atoms (Figure 4.6), then to each A it is only worth one half (just like point sharing in Chapter 1). The formula of each tetrahedron is $AX_{(4 \times 1/2)} = AX_2$.

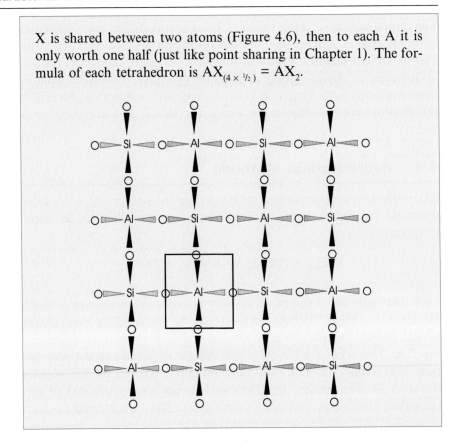

Figure 4.6 Vertex-sharing tetrahedra

The tetrahedral units link together through their vertices, using the shape of the template as a guide to form a network. For each silicon-based tetrahedral unit, *i.e.* SiO_2, there is charge balance (Si^{4+} and $2 \times O^{2-} = SiO_2$). However, for each aluminium there is a one negative charge (Al^{3+} and $2 \times O^{2-} = AlO_2^-$). Therefore, inside the network, both water molecules and counterbalancing cations are incorporated. Under different conditions one template can create a range of structures, but, in general, large templates create large pores.

Although crystallization does occur at low temperatures, hydrothermal methods are often used to speed up the reaction by raising the temperature to 160–250 °C. Numerous templating agents have been used in the formulation of zeolite materials. Microporous materials (highly crystalline, pores <12 Å) often use organic bases such as TEA (tetraethylammonium), while mesoporous materials (larger pores but lower crystallinity, pores >12 Å) often use liquid crystal templates:

$$12NaAlO_{2(aq)} + 12Na_2SiO_{3(aq)} \rightarrow Na_{12}[Al_{12}Si_{12}O_{48}].xH_2O_{(s)}$$

The symmetry of the crystal network then reflects the symmetry of the cation/anion used as a template. For example, the channel structure of the cancrinite system is shown in Figure 4.7. The three-fold symmetry of the carbonate anion is used to direct the same symmetry in the framework.

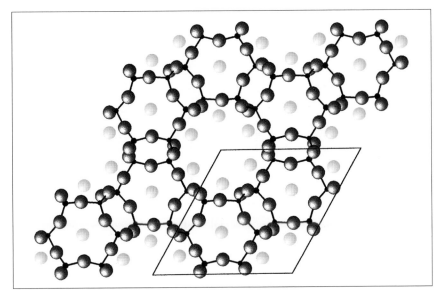

Figure 4.7 Picture of the cancrinite structure showing the three-fold axes of the framework, templated by anions with three-fold symmetry. The small black atoms are the framework tetrahedra and the light grey atoms are the framework oxygen. The orange atoms are the sodium cations, related by three-fold rotation about the centre of the large ring

Similar reactions are used in the formation of the zeotypes. They also have framework structures, but contain other elements rather than aluminium and silicon. For example, substitution of phosphorus for silicon produces the ALPO family of materials. These require no counterions in the mixture, but need mildly basic or slightly acidic conditions for crystallization. Often a small amount of hydrogen fluoride is added as a mineralizer (a material used to encourage crystallization).

Once the zeolite or zeotype is formed, the template can often be removed by heating to decompose and vaporize the organic material or by chemical treatment (*e.g.* ethanol/ethanoic acid). The pore system of the zeolite or zeotype can then be used for molecular sieving, ion exchange or catalysis.

More often than not, the solvent used for reactions of this type is water, but other solvents such as liquid ammonia and organic solvents have also been used. For example, by heating sodium silicate with ethane-1,2-diol, a framework consisting totally of silicon tetrahedra can be formed.[4]

Water heated above 100 °C (superheated) and water heated above its critical point (supercritical) are often considered too acidic for synthesis. Superheated and supercritical water are well known for their ability to destroy toxic organic chemicals, *e.g.* polychlorobiphenyls (PCBs) or chemical warfare gases (*e.g.* sarin). However, the unusual state between gas and liquid is a particularly good regime for mineral synthesis. Temperatures up to 800 °C and pressures up to 3 kbar (300 MPa) have been exploited for decades by geologists using Tuttle cold-seal bomb hydrothermal apparatus (Figure 4.8).

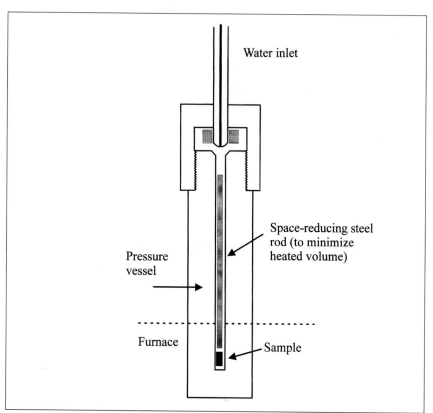

Figure 4.8 Tuttle cold-seal bomb apparatus

4.5   High Pressure

High-pressure chemistry is a somewhat expensive and dangerous activity. It can also be highly rewarding. The conditions created allow unusual oxidation states and structures to be accessed. The high pressure favours dense structures.

The simplest sort of reaction uses highly oxidized, and normally relatively unstable, compounds such as potassium chlorate. By heating potassium chlorate in a fixed volume, decomposition occurs according to the equation:

$$2KClO_{3(s)} \rightarrow 2KCl_{(s)} + 3O_{2(g)}$$

The decomposition causes a large increase in pressure (1–10 kbar, 100–1000 MPa) owing to the creation of the oxygen gas. These reactions are frequently used to generate high pressures in order produce compounds containing high oxidation states, *e.g.* PrO_2.

Beyond this range, the pressures are normally created by static pressure devices, such as piston–cylinder apparatus, solid state presses or anvil devices (opposed, tetrahedral and cubic anvil), when the sample is simply squeezed to create pressure up to 150 kbar (15 GPa) (Figure 4.9). These methods have allowed the synthesis of compounds which could not be created in any other way. However, only very small samples are produced (50 mg) in these expensive reactions, and so industrial applications of the materials produced are unlikely.

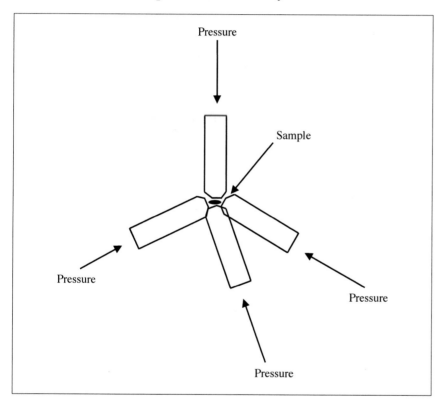

Figure 4.9 Schematic diagram of opposed tetrahedral anvil device

4.6 Single Crystals

Many of the techniques used for producing single crystals in molecular systems are simply not applicable to solid state materials. For example, solid state oxides such as the spinels (AB_2O_4) are not soluble in any solvent, at any temperature, so cannot be grown by solidification. However, there are two methods which are frequently used to produce single crystals from such materials.

4.6.1 Solidification from a Melt

There are several different ways of producing single crystals from melts.[5–8] They are all variations on a theme. The solid material is heated to slightly above its melting point in an inert crucible. Then the crystals are grown either by slow cooling or surface extrusion (see pulling method below).

Pulling Method

The material is placed in an unreactive crucible and heated to just above its melting point. A seed crystal is then dipped into the surface and slowly raised from the melt. Continuous growth then occurs at the interface.

Floating Zone

A rod of the material is held vertical and heated to cause melting at one end. By moving the heat source along the rod, melting and recrystallization occur as the source is moved. By placing a seed crystal at one end the whole rod can be converted into a single crystal.

Temperature Gradient

A large temperature gradient is created across the sample, so that crystals nucleate in the colder part of the crucible.

Flame Fusion

The solid is placed in a hopper, heated with an oxy–hydrogen torch and allowed to drip on to a seed crystal. This method is often used to grow synthetic gemstones, such as sapphires, because the other methods are not possible owing to the very high melting point of the material.

It is important to note that these methods are only applicable to solids which melt **congruently** (*i.e.* to form one phase). For multiphase systems, the **phase diagrams** must be studied to ensure that the correct stoichiometry results at the temperatures generated.

Solution Methods

For growing single crystals from solution, the methodology is similar to that used in normal hydrothermal synthesis, except that the solution is typically supersaturated and the vessels are cooled very slowly (1 °C per min) to facilitate slow growth of large crystals.

4.7   Nanomaterials

Nanomaterials are simply small clusters of material which have properties distinct from both molecular species and materials with long-range order. Typically, these clusters are formed by a sudden change of state, *e.g.* liquid → solid, consist of about 500 units and are 50–200 Å in size.

The physical properties of these materials show particular deviation from the bulk. For example, the magnetic properties of nanoparticles of iron oxide are weaker than those of the bulk material. This is particularly true of semiconducting metal chalcogenides (compounds of the heavy metals with Group 16 elements).

4.8   Amorphous Materials

Nearly all materials can be prepared in the amorphous state. The principle involves rapid quenching, which prevents the material from crystallizing in an ordered manner. This can be achieved in a number of ways, including supercooling (cooling below the freezing point), vapour deposition and gelation. The majority of these methods lead to materials which cannot be easily characterized by normal techniques such as X-ray diffraction.

Probably the most common technique for producing amorphous materials is vapour deposition, where the material is vaporized and then deposited on to a surface at a very low temperature. This is similar to the freeze drying discussed in Section 4.1. Once it is on the surface, the material can be encouraged to crystallize by heating the surface.

Very low density glasses, with very precise stoichiometries, can be synthesized using sol–gel techniques.[9] By using an organic reagent, an aerogel (a bit like expanded polystyrene) can be formed from the sol. Removal of the organic reagent then leads to an expanded glass.

4.9   Phase Diagrams and Melt Reactions

If solid state reactions are so slow, why don't we just melt the materials together and speed things up? Unfortunately, it is very rare that two materials A and B melt together and cool to form the solid AB. This process is called congruent melting and means AB (liquid) goes to AB (solid) on cooling. Most materials melt incongruently, where one of the components crystallizes first on cooling, leaving a liquid more rich in the other. Cooling of this type of material produces a mixture of A, AB and B, as shown in Figure 4.10.

In order to make sense of such problems, phase diagrams are used to investigate the partition of species at a particular temperature.

To understand the basic principles, the phase diagram for two solids

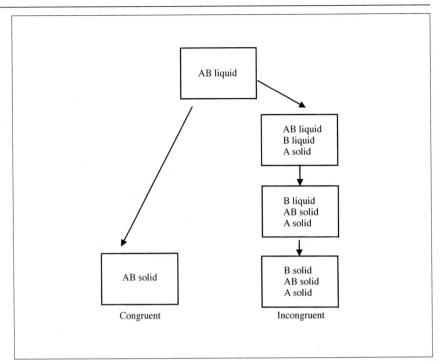

Figure 4.10 Schematic diagram of the crystallization of congruent and incongruent solids

A and B which *do not* react together is shown in Figure 4.11. Point X is the **eutectic point**, which is the lowest temperature at which the whole system is molten. The curved line separating the solid and liquid region is called the **liquidus**. The *x* axis simply represents the ratio of the two components. As A and B do not react, there are no vertical lines within the diagram.

Figure 4.12 represents two different cases where A and B *do* react to form AB. Part (i) shows a phase diagram for a compound AB which

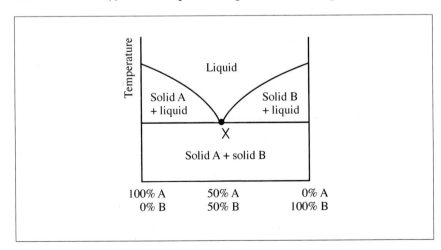

Figure 4.11 Phase diagram for A and B, which do not react together

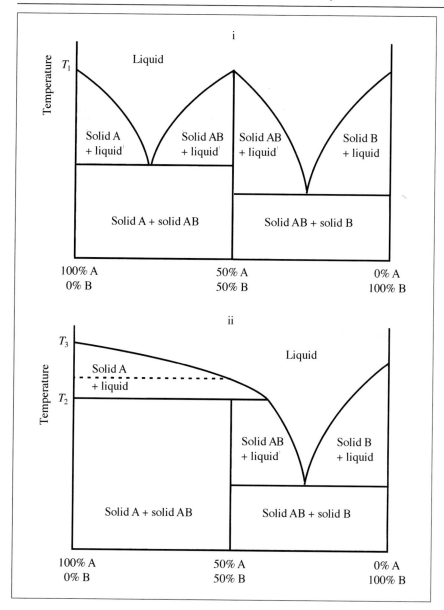

Figure 4.12 Phase diagrams for solids A and B which (i) melt congruently, (ii) melt incongruently

melts congruently. Raising the temperature beyond T_1, with equal proportions of A and B, will lead to solid AB on cooling. Obviously, moving to the right would produce a mixture of AB and B. In an analogous diagram for an incongruently melting material, part (ii), cooling a 50:50 mixture of AB from the liquid state first deposits solid A (below T_3), and then beyond T_2 starts depositing AB. However, now the liquid is rich in B, so at any reasonable rate of cooling, a mixture of solid products, A, AB and B, is obtained. In these cases it is advisable to keep below T_2, known as the peritectic temperature. This means that incongruently melting mixtures cannot be allowed to melt and the synthesis must be performed in the solid state.

Summary of Key Points

1. *The majority of solid materials* are prepared by solid state reaction of the appropriate constituents at elevated temperature. Suitable choice of starting materials and their purity are imperative in order to produce good quality products.

2. *The use of special atmospheres* (such as hydrogen) or vessels (such as sealed tubes) can be used to access unusual oxidation states, or to produce air-sensitive materials.

3. *Precursor methods*, such as co-precipitation, can be used to generate smaller particle sizes and allow reactions to proceed faster or at lower temperature.

4. *Hydrothermal methods* can be used to form solid state materials, providing the reactants are soluble. Slow cooling from solution can be one way to form single crystals.

5. *Phase diagrams* can be used to identify likely preparative temperature/state conditions for materials synthesis.

Further Reading

M. T. Weller, *Inorganic Materials Chemistry*, Oxford University Press, Oxford, 1994.
C. N. R. Rao and J. Gopalakrishnam, *New Directions in Solid State Chemistry*, Cambridge University Prerss, Cambridge, 1999.
A. R. West, *Basic Solid State Chemistry*, Wiley, New York, 1997.

Problems

1. What would be suitable starting materials for the preparation of the superconductor $YBa_2Cu_3O_7$?

2. Preparation of cobalt aluminate from CoO and Al_2O_3 requires a reaction temperature of 1200 °C. How could the reaction temperature be lowered?

3. How could a pure sample of the compound Li_2TiO_3 be prepared from a mixture of lithium oxide and titanium dioxide when the lithium oxide is hygroscopic?

4. How could a sample of V_2O_3 be prepared from a mixture of V_2O_5 and V (metal)?

5. If a sample of cobalt aluminate was prepared as in question (2) via the high-temperature method, what would be a suitable choice of reaction vessel?

6. If a 20:80 mixture of A and B was cooled to room temperature in the phase diagram in Figure 4.10(ii), what solids would precipitate?

7. How could a single crystal of a zeolite be prepared for structure determination?

References

1. P. F. Henry and M. T. Weller, *Angew. Chem., Int. Ed. Engl.,* 1998, **37**, 2855.
2. S. E. Dann, D. B. Currie and M. T. Weller, *J. Solid State Chem.,* 1992, **97**, 179.
3. S. Mintova, N. H. Olsen, V. Valtchev and T. Bein, *Science,* 1999, **283**, 958.
4. D. M. Bibby and M. P. Dale, *Nature,* 1985, **317**, 157.
5. R. A. Laudise, *The Growth of Single Crystals,* Prentice-Hall, Englewood Cliffs, NJ, 1970.
6. J. M. Honig and C. N. R. Rao, *Preparation and Characterization of Materials,* Academic Press, New York, 1981.
7. E. Banks and A. Wold, *Solid State Chemistry,* Dekker, New York, 1974.
8. S Mroczkowski, *J. Chem. Educ.,* 1980, **57**, 537.
9. D. J. Hamilton and C. M. B. Henderson, *Silicate Synthesis,* 1968, 832.

5

Electronic and Magnetic Behaviour of Solids

Many of the industrial applications of materials make use of their particular electronic and magnetic properties. These properties are often directly related to their physical structure. Since the atoms/ions in the extended structure of these materials are very close together, the interactions between them occur *throughout* the lattice.

The idea of metallic conductivity should now be familiar. Electrons are the charge carriers which move through a lattice consisting of metallic ions. The term 'metallic conductivity' is somewhat confusing, as this behaviour is not limited to metals and alloys. It is also shown by many oxide and sulfide materials. Two other types of conductors will also be discussed in this chapter: superconductors, where electrons move cooperatively (possibly as pairs), and ionic conductors, where the charge carriers are moving ions.

Aims

By the end of this chapter you should be able to:

* Use band theory to explain the differences in electronic properties of conductors and semiconductors with temperature
* Understand the differences between electronic and ionic conductivity
* Explain the differences between paramagnetism, ferromagnetism and antiferromagnetism
* Understand the similarities and differences between conductivity and superconductivity

Finally, different sorts of magnetic behaviour will be discussed. Initially electron spins acting discretely in molecules and lattices will be compared, before we move on to different types of magnetic ordering mechanism.

5.1   Useful Properties of Solids

Many of the useful properties of solids result from their physical characteristics. For example:

1. Malleable metals and alloys are used for making implements
2. Diamonds are used for cutting owing to their physical hardness
3. Alumina is used for crucibles owing to its high temperature stability and thermal shock resistance

However, a large number of materials also find uses owing to the behaviour of some of the electrons within them. For example:

4. Copper is used extensively in electrical circuitry owing to its low cost and high conductivity
5. Ultramarine is a pigment which contains the S_3^- anion as part of its structure. Electronic transitions in the anion, caused by visible light, give a strong blue colouration
6. Barium ferrite is used in permanent magnets owing to the strong magnetic interaction of the unpaired electrons of the iron ions in the lattice

These properties can be understood by considering how the ordered array of atoms/ions in a lattice causes interaction between the electrons in the orbitals on the *individual* lattice points.

5.2   Electronic Behaviour of Solids

The combination of atomic orbitals to form molecular orbitals can be used to describe the properties of molecules. For example, the two unpaired electrons in the oxygen molecule, O_2, can be explained by molecular orbital theory. Simple electron sharing cannot account for these unpaired electrons (Figure 5.1). This approach can also be extended to infinite lattices, where n atomic orbitals create n molecular orbitals, divided equally between bonding and antibonding combinations. The bonding and antibonding levels are so closely spaced that they appear *continuous*. This arrangement generates certain characteristic properties, depending on whether the material is conducting or semiconducting.

The differences in conductivity of conducting and semiconducting materials as a function of temperature is shown in Figure 5.2. Two observations follow from the graph:

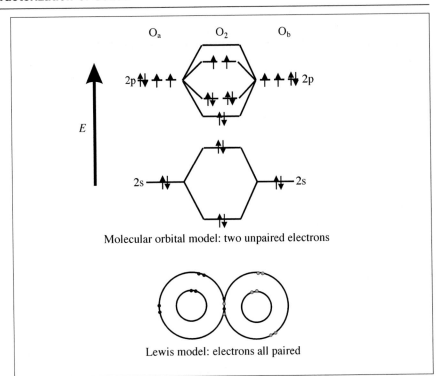

Figure 5.1 Comparison between molecular orbital model and Lewis model for the bonding of O_2

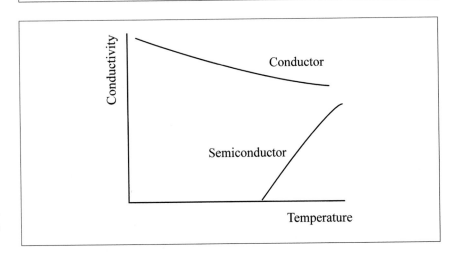

Figure 5.2 Conductivity of semiconductors and conductors as a function of temperature

Conductor: the curve indicates that the conductivity falls off with increasing temperature, but persists down to very low temperature.

Semiconductor: a semiconductor is a substance with an electric conductivity that increases with increasing temperature. Below a particular temperature the material is no longer conducting. This suggests that a minimum temperature is required before the electrons or ions are given enough energy to move.

Worked Problems

Q Why does the resistance of a metal increase with temperature?

A As the temperature rises, the atoms or ions in the lattice are given more thermal energy. This energy would normally be shared between rotation, translation and vibration. In the lattice the atom or ion is *fixed* on its lattice site and cannot rotate or translate. Therefore, the vibration of the atom or ion increases dramatically with temperature, effectively making the cross-section of the atom bigger (Figure 5.3). The electrons or ions passing through the lattice have a greater chance of colliding with the 'larger' atoms and not passing through the lattice. This means the resistance of the conductor gradually increases with temperature.

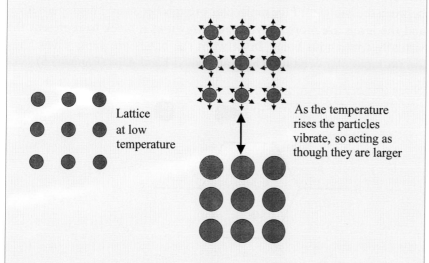

Lattice at low temperature

As the temperature rises the particles vibrate, so acting as though they are larger

Figure 5.3 Increase in particle cross-section as temperature rises

Q Why have insulators not been included on the graph in Figure 5.2?

A Insulators have not been included on the graph as they are just a special sort of semiconductor and would have a similar shaped graph. The temperature required to give the conducting species enough energy to move in an insulator is just much greater than for a semiconducting material.

As discussed in Chapter 1, the fundamental properties of a metallic lattice can be described by the process of each atom contributing one or more electrons to a delocalized sea of electrons. This concept can be explained more fully using an extension of molecular orbital theory known as band theory.

5.2.1 Band Theory

In a diatomic molecule formed from two atoms, an atomic orbital on one atom combines with an atomic orbital on the another atom, providing the orbitals have the correct symmetry. They generate bonding and antibonding combinations, as shown in Figure 5.4. Each atomic orbital creates one molecular orbital. Extending this to 2, 8 or 32 atoms generates 2, 8 or 32 molecular orbitals (Figure 5.5). Every pair of orbitals creates a bonding and antibonding combination. With increasing numbers of atoms, the separation between the levels above (antibonding) and below (bonding) the non-bonding line becomes smaller and smaller. For an infinite number of atoms (as in a solid state lattice), the difference between the energies of the molecular orbitals becomes indistinguishable, and the levels are more conveniently drawn as a block to represent the virtually continuous band. In between the bands are gaps, where they are no molecular orbitals. These are called band gaps (Figure 5.6).

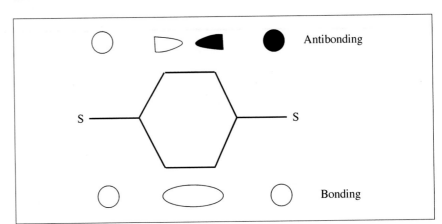

Figure 5.4 Schematic representation of the formation of bonding and antibonding combinations of two s-orbitals

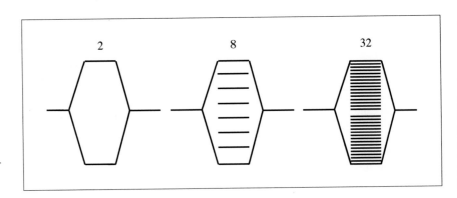

Figure 5.5 Formation of n molecular orbitals from n atomic orbitals

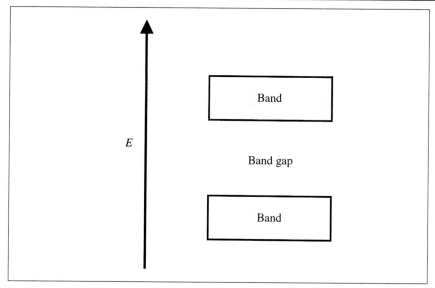

Figure 5.6 Bands and band gaps

Providing the atomic orbitals are large enough, each set of orbitals will create a band by overlap with the orbitals on neighbouring atoms in the lattice. These can be derived from filled and unfilled levels. For example, sodium has the electronic structure $1s^2 2s^2 2p^6 3s^1$. The first two shells of the sodium are held close to the nucleus and do not extend out far enough from the lattice point to interact with the orbitals of the atoms on neighbouring lattice points. However, the 3s (partially filled) and 3p (empty) orbitals are sufficiently large enough in size to overlap with the corresponding orbitals on neighbouring atoms to produce a band.

The outer electrons from the atoms are then placed in the bands in a similar way to molecular orbital theory. This electronic arrangement leads to different properties, depending on the level of filling of the bands. The highest energy band containing electrons is referred to as the valence band as it contains the outer (valence) electrons. The lowest energy empty band above the valence band is called the conduction band. The separation between the valence and conduction bands dictates the overall electronic properties.

For conduction to occur, an electron must be able to move to a new energy level. A material with a filled valence band and a large energy gap between this level and the conduction band is an insulator, since there is a large energy barrier for movement of an electron to a new level.

The concentration of energy levels within a band is called the density of states; this is given the symbol $N(E)$, and is determined for a small increment of energy dE. Figure 5.7 shows three different band diagrams. The shaded portion of the band is occupied and the unshaded portion of the band is empty. The width of the band depends on the degree of overlap between the atomic orbitals.

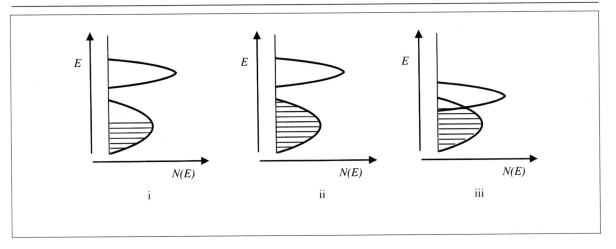

Figure 5.7 Density of states diagrams for (i) and (iii) conductors and (ii) semiconductor/insulator

Worked Problem

Q The band structures of two solids are shown in Figure 5.8. Which band structure is indicative of good overlap?

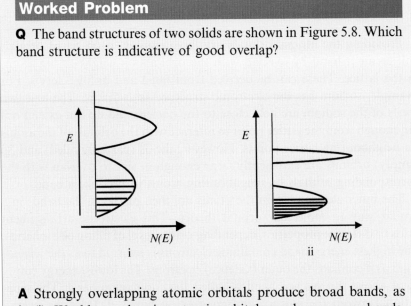

A Strongly overlapping atomic orbitals produce broad bands, as in (i). Weakly overlapping atomic orbitals produce narrow bands. For this reason, for orbitals with the same principal quantum number (main energy level), the s bands tend to be broader than p, owing to better overlap.

Figure 5.8 Density of state diagrams for good and poor overlap

The separation of the bands depends on the original separation of the atomic orbitals. If the original atomic orbitals were very close in energy, the bands will also be very close in energy. If the overlap of the orbitals is good and the original orbitals were close in energy, the bands can overlap as shown in Figure 5.7(iii).

The change in conductivity with temperature in Figure 5.2, for a conductor and semiconductor, can now be explained. For conduction to occur, the electrons must have enough energy to move to a vacancy in another energy level. For this reason, conduction does not occur in filled bands as the electrons have no vacancy to move to. The energy separation between the valence and conduction levels in a conductor is very small, as the bands are overlapping or partially filled. This means very little energy is required to move an electron to a new level where it can move. Movement of an electron also results in a vacancy (or hole) in the level it has just moved from. In a semiconductor, the band gap *between* bands is significant but small, typically 1–3 eV. At low temperature, there is not enough energy to overcome this gap and the material is not conducting. However, as the temperature rises the electrons gain enough energy to jump the gap and the material becomes conducting. An insulating material would have a band gap of much greater than 3 eV (*e.g.* for diamond the band gap is 6 eV). There is not enough energy to jump this gap at practical temperatures which would allow the material to be used as a semiconductor.

Table 5.1 Typical band gaps (eV) of insulators and semiconductors

Insulators		Semiconductors	
C (diamond)	6	Si	1
NaCl	9	Ge	0.7
NaBr	8	Sn (grey)	0.1
KF	11	GaAs	1.4
KCl	9	InP	1.3

Materials which naturally have a small band gap, like silicon and germanium, are called intrinisic semiconductors. Examples of some intrinsic semiconductors with their band gaps are given in Table 5.1. Elements and compounds which are not semiconducting in the pure state can be made semiconducting by doping. These materials are called extrinisic semiconductors. Doping involves replacing a very small amount of the element in the lattice with another element with either fewer or more electrons per atom. The doping element must have energy levels similar to those of the host. Figure 5.9 shows the effect of doping silicon, with electronic configuration $1s^2 2s^2 2p^6 3s^2 3p^2$, with aluminium (one less electron) and phosphorus (one more electron). The valence band of phosphorus creates a donor band, which lies at higher energy than the valence band of the silicon host (owing to the higher effective nuclear charge). The donor band is closer in energy to the conduction band of silicon than the silicon valence band. Hence, promotion of an electron from the donor band to the conduction band in the doped material requires less energy, and will occur at a lower temperature than in the pure material. This process creates a 'n-type' extrinisic semiconductor, where the n stands for *negative*. This is because the charge carriers added by the phosphorus are electrons which are negatively charged. The opposite is true for addition of aluminium. Addition of very small amounts of aluminium creates a conduction band lower in energy than the host silicon lattice. For each aluminium, a hole (a vacancy for an electron) is added. These holes can accept an electron from the valence band, again decreasing the energy barrier for conduction. This type of extrinsic semiconductor is called 'p-type', where p stands for *positive*, because the charge carriers added by the aluminium are positive holes.

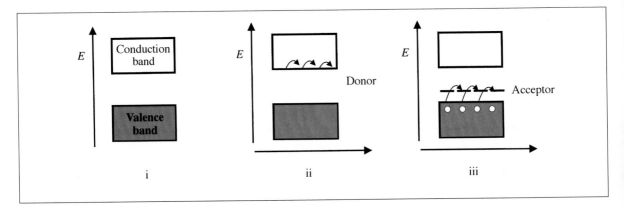

Figure 5.9 Effect on the band structure of doping silicon (i) with phosphorus (ii) and aluminium (iii)

Worked Problem

Q Why does the aluminium conduction band lie at lower energy than the silicon conduction band?

A Aluminium lies to the left of silicon in the Periodic Table. Its atom has one fewer proton and so the nuclear charge is more effectively screened. The conduction band of aluminium will, therefore, lie at lower energy.

At $T = 0$ K the electrons sequentially fill the molecular orbitals according to the Aufbau principle (two at a time, with opposing spin). The level to which a band structure is filled is call the Fermi level. If the Fermi level lies within a band the material is expected to be conducting. If the Fermi level lies at the top of a band, it is normally drawn midway between this band and the next (Figure 5.10). In this case the material is semiconducting or insulating.

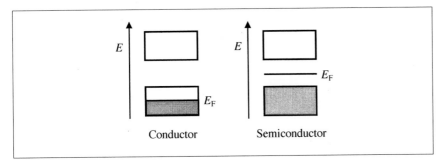

Figure 5.10 Fermi levels of a conductor and semiconductor

At temperature above $T = 0$ K the population (P) of the orbitals has a Boltzmann-type distribution which is described by the **Fermi–Dirac** distribution function:

$$P = \frac{1}{e^{(E - \mu) / kT} + 1}$$

where μ is the **chemical potential** (the energy level for which $P = \frac{1}{2}$).

Band Theory: A Case Study of the First-row Transition Metal Oxides

Band theory can successfully be used to explain the marked difference in conduction behaviour of the oxides of the early and late transition metals of the first period. The first-row transition metal monoxides form the halite structure as shown in Figure 5.11. Each metal ion is in

Figure 5.11 Sodium chloride structure of the first-row transition metal monoxides

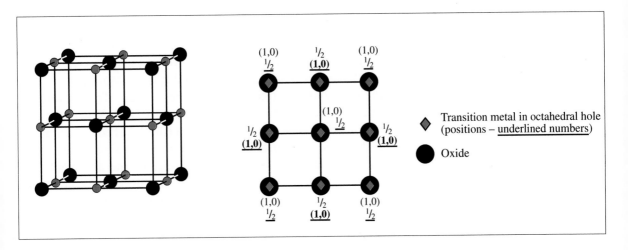

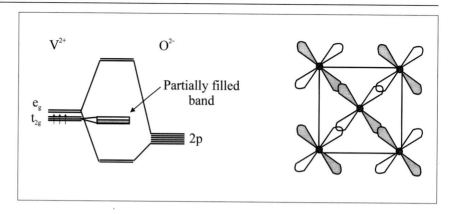

Figure 5.12 Overlap of t_{2g} orbitals to form a band for an early transition metal monoxide

an octahedral ligand field which splits the d-orbitals into t_{2g} and e_g levels. The close proximity of the metal atoms mean that these levels should overlap and form bands (Figure 5.12).

The differences in conductivity of the transition metal monoxides can easily be observed from their colour. Nickel(II) oxide is green, which is the same colour expected for a discrete nickel(II) complex in solution. Visible light of a specific wavelength is absorbed, corresponding to the difference between two well-separated energy levels. This indicates that the nickel ions within the structure are behaving discretely, just like they would do in solution. In contrast, vanadium(II) oxide is black. When a material appears black to the eye it is absorbing light over the full visible range, implying that there are a lot of energy levels close together. This is the colour expected for a conductor with a band structure, where the vanadium ions are acting cooperatively.

These differences can be explained by considering the different location of the valence d-electrons. For the vanadium, the valence electrons are in the t_{2g} orbitals. These orbitals point towards each other and can overlap to form a band. This creates a partially filled band and hence a good conductor, as expected from Figure 5.12. In contrast, the unpaired valence electrons for nickel are in the e_g level. These orbitals are separated from each other by oxygen atoms, which prevents good overlap. These orbitals remain localized on the nickel ions, as shown in Figure 5.13. Hence in nickel oxide the atoms behave discretely, and it is an insulator.

Band theory should be treated with some caution. The presence of bands in solid state structures is well supported by X-ray emission and absorption data where energy is emitted (or absorbed) over a range relating to the band structure. However, even in this simple case, band theory fails to explain why MnO is insulating. In the argument used above, MnO has electron vacancies in the t_{2g} and should be conducting.

As a general guide, the following rules can be used to predict whether overlap is *likely* to be good between the d-orbitals of transition metal compounds:

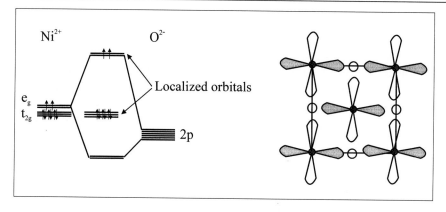

Figure 5.13 Discrete molecular orbitals of nickel(II) oxide

1. The formal charge on the cation should be small
2. The cation should occur early in the transition series
3. The cation should be in the second (or third) transition series
4. The difference in electronegativity between the cation and anion should be small (for the same reasons noted for ionic solids in Section 5.2.3)

5.2.2 Superconductivity

Superconductors are a special sort of material which conducts with no resistance below a temperature called the critical temperature (Figure 5.14). As we have seen, the conductivity of conductors falls with increasing temperature, as the lattice vibrations increase and hinder the flow of current. In a superconductor the electrons are thought to move in pairs *with* the lattice vibrations in a concerted process, so there is no resistance. Bardeen, Cooper and Schrieffer[1] proposed a theory (called BCS theory) to explain superconductivity involving the movement of electrons with *opposite spin* in *pairs*. When an electron moves past a lattice point,

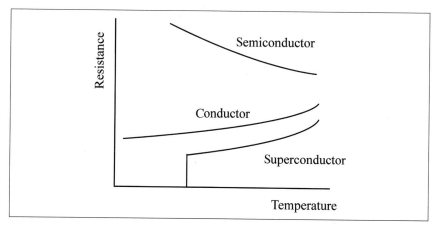

Figure 5.14 Resistance of conductors, semiconductors and superconductors as a function of temperature

the nearest positively charged atom is drawn towards it slightly. The local distortion of the lattice creates an area rich in positive charge which a second electron is attracted towards. This effect continues throughout the lattice, allowing movement of the electrons in pairs. Therefore, the weak lattice vibrations which occur at low temperature assist the superconducting mechanism. However, above the critical temperature, the lattice vibrations become too strong and overcome the slight attraction between the electrons, and the superconductivity is lost.

5.2.3   Ionic Conductivity

Ionic compounds, like sodium chloride, are normally considered insulating in the solid state. This is due to the marked difference in electronegativity of the cations and anions, which creates a large band gap of typically 6–12 eV between the valence and conduction bands. Therefore, ionic solids have a band structure similar to an insulator (Figure 5.15). The ions and their associated electrons can be thought of as fixed on their lattice sites.

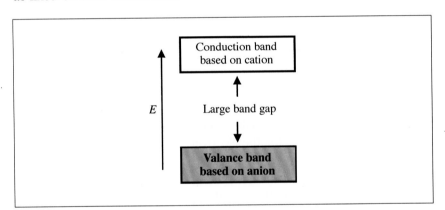

Figure 5.15 Band structure of sodium chloride

Worked Problem

Q Would you expect orbitals based on the sodium cation or the chloride anion to be larger?

A The chloride anion is negatively charged, and owing to the greater electrostatic repulsion between electrons the orbitals will be larger and lie at lower energy.

However, most lattices are not totally perfect, as will be seen in Chapter 6. There are often ion vacancies in the lattice. At high temperature, often close to the melting point, ionic solids can become conducting. The thermal treatment gives the ions enough energy to move from their lattice sites into a vacancy, as shown schematically in Figure 5.16. Some typical conductivities for electronic and ionic conductors are given in Table 5.2. It is apparent that ionic conductivities in solids are *many orders of magnitude* less than electronic conductivities.

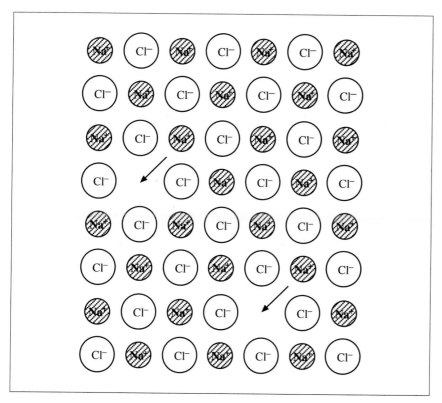

Figure 5.16 Schematic diagram of ionic conduction in sodium chloride

Table 5.2 Typical electrical conductivities

Type of material	Electronic ($\Omega^{-1}\ cm^{-1}$)	Ionic ($\Omega^{-1}\ cm^{-1}$)
Metal	10^{-1}–10^5	–
Semiconductor	10^{-5}–10^2	–
Insulator	$<10^{-10}$	–
Ionic solids, *e.g.* NaCl	–	10^{-18}–10^{-4}
Fast ion conductor	–	10^{-3}–10^1
Molten salts	–	10^{-3}–10^1

Conduction can occur by ions either moving to an empty lattice site or moving to an interstitial site. An interstitial site is a gap in the lattice which is large enough to accept an ion, but is not a normal lattice site. This type of behaviour will be discussed further in Chapter 6.

Solid Electrolytes or Fast Ion Conductors

A small number of solids have an unusual structure, where only one set of ions is fixed on its lattice sites. The other ions are *free to move* through the lattice. These materials are intermediate between the ionic solids (where all the ions are fixed on their sites) and a liquid electrolyte (where both sets of ions move).

Ionic conductors do not pass the same level of current as normal electronic conductors. However, as they are often very high melting point ceramic materials (1500 °C), they can be used at temperatures at which metallic conductors would melt. Typically, they consist of non-stoichiometric materials (Chapter 4) with open tunnels or layers containing vacant sites. Conduction occurs by the ions moving between the sites which are not totally filled.

As previously discussed, ions on fixed lattice sites normally only vibrate and do not translate or rotate. In the fast ion conductors this is not the case: one set of ions *can* migrate. This type of behaviour normally only occurs at high temperature, and is more common with cations than anions, owing to the smaller size of the cations. It is noteworthy that this phenomenon is therefore often linked to only one particularly high temperature polymorph of a material. For example, silver iodide is practically insulating at room temperature but at high temperature becomes conducting. This is due to a structural change to the α-AgI polymorph, which has an open structure in which the silver ions can move.

One extremely important member of this class is sodium β-alumina. β-Alumina has a structure based on close-packed layers of oxygen with ordered oxygen vacancies. Sodium cations reside in the vacancies created by the oxygen deficiency and migrate from site to site. This process is particularly easy, as there are many more sites available than sodium atoms and the sodium cation is small compared with the oxygen anion. Schematically this could be represented as shown in Figure 5.17

Anion conductors are more unusual owing to the much larger size of anions. Most are a result of defect lattices and in many cases these materials have structures related to the fluorite lattice (Chapter 1); for example, the high-temperature polymorphs of PbF_2 or ZrO_2. Anion vacancies on the lattice points increase with temperature and give more empty sites to which the anions can move. The cubic polymorph of zirconia, with the fluorite structure, can be stabilized at lower temperature by using dopants to produce stabilized zirconias. The introduction of calcium or yttrium on the zirconium site creates anion vacancies for charge balance.

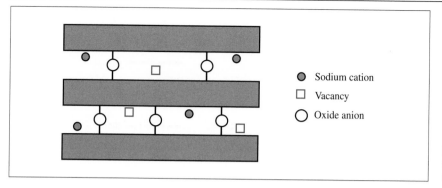

Worked Problem

Q Would ionic conductors or electronic conductors have the higher conductivity?

A Electronic conductors. Ionic conductivity is much lower as the ions are much larger. This means ions require more energy to move through the lattice and overcome energy losses, such as collision with the fixed lattice ions.

In principle, a whole range of applications can be envisaged using these solid electrochemical cells in situations where cells containing liquids could not function, *e.g.* some fuel cells.

5.3   Magnetic Behaviour

In a similar way to their electronic properties, the magnetic properties of three-dimensional solids result from the interaction of the metal centres. For example, transition metals have partially filled d-orbitals and interactions of the electrons in these orbitals can lead to magnetic behaviour.

In a discrete molecule in solution, the unpaired electrons act totally independently and are randomly oriented. For example, divalent nickel in both an octahedral and square-planar ligand field is shown in Figure 5.18. Random orientation of the unpaired electrons and their associated magnetic fields, as in the case of octahedral nickel, is known as paramagnetism. When all the electrons are paired, as in the case of the square-planar nickel complex, the material is called diamagnetic.

In an ordered solid at high temperature, or when the metal centres are far apart, the electrons will also behave paramagnetically. The change in magnetic moment of a material with temperature is called its mag-

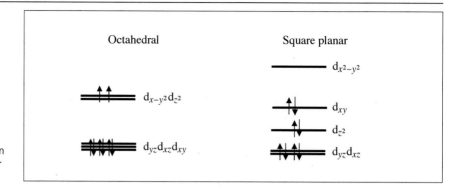

Figure 5.18 Crystal field diagram for octahedral and square-planar nickel(II)

netic susceptibility. The magnetic susceptibility graph for a paramagnetic solid is shown in Figure 5.19. As the temperature is lowered, thermal effects can no longer overcome the ordering of the electrons and the level of magnetization rises. This behaviour is adopted by the majority of compounds at high temperature, where thermal disordering overcomes any magnetic interaction between the metal centres. The magnetic susceptibility for a paramagnetic compound is inversely proportional to the temperature, as described by Curie's law:

$$\chi_{\mathrm{M}} = \frac{C}{T} \quad \text{where } C \text{ is a constant}$$

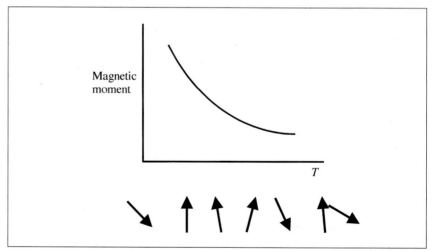

Figure 5.19 Change in magnetic susceptibility (χ) with temperature (T)

In a solid lattice, where the unpaired electrons on neighbouring ions are close together, the spins can couple together and lead to a number of magnetic ordering mechanisms, as summarized in Table 5.3. Interaction of these spins leads to a modification of the Curie law to include the temperature at which these mechanisms occur, θ. This is called the Curie–Weiss law:

$$\chi_M = \frac{C}{T - \theta}$$

Table 5.3 Field and temperature dependence of magnetism

Magnetism	Effect of external magnetic field	Temperature dependence	Field dependence
Diamagnetism	Weak repulsive	None	None
Paramagnetism	Weak attractive	$1/T$	None
Ferromagnetism	Strong attractive ⎱	Complex field and temperature	
Antiferromagnetism	Weak attractive ⎰	dependency	

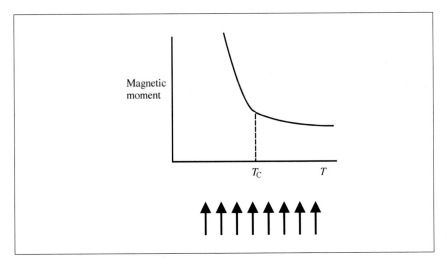

Figure 5.20 Magnetic susceptibility curve for a ferromagnetic material

5.3.1 Ferromagnetism

If all the unpaired electrons align themselves parallel with the external magnetic field, the magnetic material is said to be ferromagnetic. This leads to the magnetic susceptibility graph shown in Figure 5.20. A sharp rise occurs in the magnetization when the electrons align with the field. The temperature at which this occurs is called the Curie temperature (T_C).

5.3.2 Antiferromagnetism

If the magnetic spins on neighbouring atoms align antiparallel, the material is described as antiferromagnetic and there is an overall drop in the magnetization. This behaviour shows itself in a sudden drop in the mag-

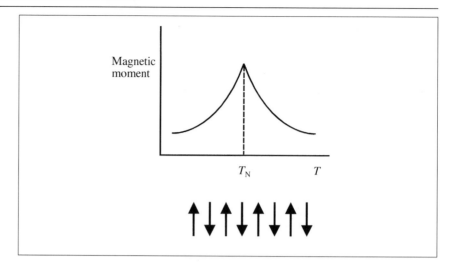

Figure 5.21 Magnetic suscepti-bility curve for an antiferromag-netic material

netic susceptibility at a particular temperature, as shown in Figure 5.21. The temperature at which this occurs is called the **Néel temperature** (T_N).

5.3.3 Ferrimagnetism

The structure of a spinel and the magnetic moments of its electrons is shown in Figure 5.22. Spinels such as $CoFe_2O_4$ contain two magnetic ions on two different sites, one octahedral and one tetrahedral in the ratio 2:1. When these materials order antiferromagnetically, there is an

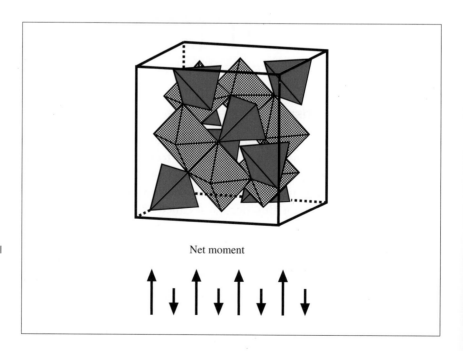

Figure 5.22 Structure of a spinel and ordering of its magnetic moments. Larger arrows repre-sent more electrons than do smaller ones, giving an overall moment despite the antiferro-magnetic ordering

overall positive magnetic moment as there are more unpaired electrons on one site than on the other.

These three magnetic ordering mechanisms are the simplest and rely on the 180° pairing of spins. This is not necessarily always the case. For example, a canted antiferromagnet has its spins aligned *at an angle* to the field. The arrangement can be quite complex, leading to a magnetic unit cell much bigger than the nuclear unit cell, with unusual structures, *e.g.* in the shape of a helix.

5.3.4 Strength of Magnetic Interactions

The magnetic ordering temperatures (T_C and T_N) are generally dependent on a number of factors, some of which are *directional*. Therefore, magnetic ordering may not occur simultaneously in all directions. The following lists some of the factors which affect the magnetic ordering temperature.

Number of Unpaired Electrons

As the number of unpaired electrons increases, the magnetic ordering temperature rises. For example, in Table 5.4,[4] as the number of d-electrons increases the magnetic ordering temperature rises.

Table 5.4 Effect of number of unpaired electrons on magnetic ordering temperature for $Sr_2MO_4^{2-}$

M	V	Cr	Fe
Number of unpaired d-electrons in M^{4+}	1	2	4
Ordering temperature (K)	10	35	60

Distance between the Interacting Metal Centres

In a way similar to holding bar magnets close together, the closer the distance the stronger is the interaction. This can lead to materials which magnetically order in two dimensions at much higher temperature than they order in three dimensions. For example, Figure 5.23 shows a double layer of magnetic centres which are closely separated in two dimensions but are much further apart in the third. This type of structure can lead to different magnetic ordering mechanisms in the different directions. For example, $Rb_3M_2F_7$ (M = Co, Mn)[5,6] orders antiferromagnetically *in the layers* but ferromagnetically *between them*.

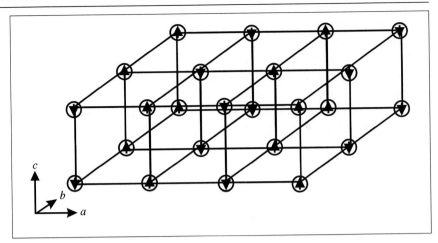

Figure 5.23 Ferromagnetic ordering between the layers and antiferromagnetic ordering in the layers of a bidimensional magnet

5.3.5 Superexchange Mechanism

The transition metal monoxides, *e.g.* NiO, FeO and CoO, magnetically order by a process known as 'superexchange'. This process involves interaction of the transition metal d-electrons with the oxygen p-electrons, resulting in an antiferromagnetic arrangement of the spins on the transition metal ions (Figure 5.24).

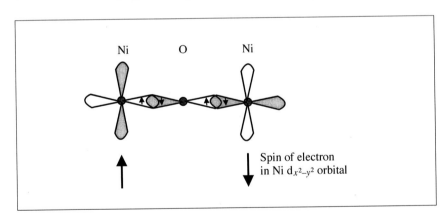

Spin of electron in Ni $d_{x^2-y^2}$ orbital

Figure 5.24 Superexchange mechanism

In NiO, for example, divalent nickel has eight d-electrons, which gives rise to two unpaired electrons in the e_g set for octahedral coordination. It crystallizes with the sodium chloride structure, where the nickel d-orbitals overlap with the oxide p-orbitals along each cell edge. For example, each nickel has one electron in the d_{z^2} orbital which can interact with the electron in the oxygen p_z orbital and form an antiparallel spin pair. The other electron in the oxygen p_z orbital must be antiparallel to that, as it is in the same orbital as the other oxygen electron and each orbital can contain only two electrons of opposite spin. The other electron in the oxygen p_z in turn interacts with the next electron in a d_{z^2} orbital on the next nickel, which must be antiparallel. This creates a chain of antiparallel spins running through the structure. This process occurs simultaneously in all three directions, and gives a magnetic lattice as shown in Figure 5.25. The magnetic unit cell must have an identical electron with the same spin on each corner (i.e the definition of a unit cell is the *smallest repeating unit*, which shows the *full symmetry* of the system). The magnetic unit cell is therefore $\sqrt{2}$ times the nuclear cell parameter, as shown in the diagram.

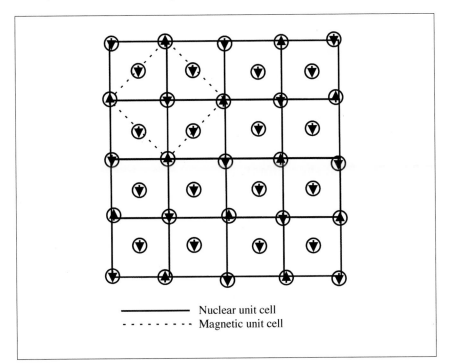

——————— Nuclear unit cell
- - - - - - - - - Magnetic unit cell

Figure 5.25 Nuclear and magnetic unit cell for nickel oxide

5.3.6 Magnetic Frustration

Magnetic frustration was first recognized in sodium titanate, $NaTiO_2$. It results from triangularly related lattice points which house magnetic ions as shown in Figure 5.26. A typical antiferromagnet would rotate freely to adjust the neighbouring spins, so they are exactly antiparallel to each other. The arrangement (topology) of the atoms in this material, with triangular arrays of magnetic ions, means the pairwise exchange cannot be satisfied for all three pairs at once. The lattice is described as *frustrated*, and normally a compromise is reached where the spins lie at 120° to each other.

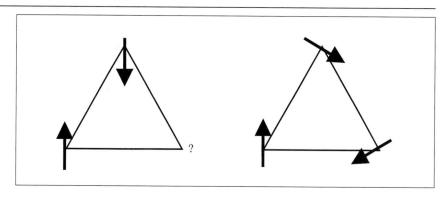

Figure 5.26 Magnetic frustration

5.3.7 Spin Glass

A spin glass is a material which below a particular temperature locks its spins into a random orientation. This means there is no regular ordering, and the spins just seem to have frozen in a way reminiscent of a glass (hence the name).

5.3.8 The Meissner Effect

Magnetism was believed to destroy superconductivity until very recently. Superconductors have a unique way of preventing magnetic flux penetrating the sample. When a magnet is brought close to a superconductor below the critical temperature, a current of electrons is set up on the surface (Figure 5.27). As a moving charge creates a magnetic field, this exactly opposes the external magnetic field and prevents the external magnetic flux penetrating the sample. This is known as the Meissner effect. This effect is observed as repulsion of a magnet, and is strong enough to cause a magnet to levitate above the superconductor. This effect is dependent on the applied field and can occur in two different ways:

Figure 5.27 Exclusion of magnetic field from a superconductor

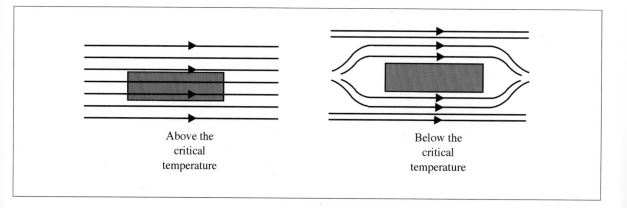

Above the
critical
temperature

Below the
critical
temperature

Type 1: The transition between the normal state and the superconducting state is a one-step process, and the material is either perfectly diamagnetic or not at a particular applied field.

Type 2: This behaviour is not so clear cut. At small values of the applied field, the material behaves in the same way as a type 1 superconductor and there is not penetration by the field. Similarly, at high values of the applied field the field readily penetrates the whole sample. However, at intermediate values, between the two extremes, there is partial penetration by the field and the sample exhibits a complex structure. There are mixed regions in the superconducting and normal state. This is known as the **vortex state**. This means that in type 2 materials the magnetization diminishes gradually rather than suddenly (Figure 5.28)

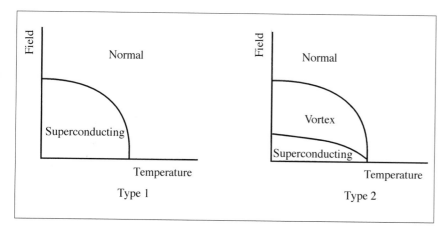

Figure 5.28 Field and temperature dependence of type-1 and type-2 superconductivity

Summary of Key Points

1. Band Theory
The differences in properties between semiconductors and conductors explained by the interaction of the whole lattice. Intrinsic and extrinsic semiconductors and donor/acceptor bands. Simple example of band theory in action.

2. Ionic Conductivity
Differences between normal ionic solids and ionic conductors. Temperature dependence of conductivity.

3. Magnetic Ordering Mechanisms
Paramagnetism, diamagnetism, ferromagnetism and antiferromag-

netism. Variation in magnetic susceptibility with temperature and strength of magnetic interaction. Superexchange mechanism.

4. Unusual Magnetism
Superconductivity, magnetism and the Meissner effect. Magnetism frustration and spin glasses.

Further Reading

A. R. West, *Basic Solid State Chemistry*, Wiley, New York, 1997.
D. W. Bruce and D. O'Hare, *Inorganic Materials*, Wiley, New York, 1992.
M. T. Weller, *Inorganic Materials Chemistry*, Oxford University Press, Oxford, 1994.
J. E. Huheey, E. A. Keiter and R. L. Keiter, *Inorganic Chemistry*, 4th edn., Harper/Collins, New York, 1993.

Problems

1. Draw the band structure expected for germanium doped with gallium. What sort of semiconductor would this be? What type of charge carrier has been added?

2. Graphite has a layered structure consisting of fused six-membered rings and is conducting at reasonable temperatures. Why would you expect diamond to be insulating?

3. A p-type semiconductor can be created by increasing the amount of anion in the formula ZnO. Explain.

4. Can the following be used to differentiate between a semiconductor and an insulator?
(a) The shape of the conductivity graph versus temperature.
(b) The temperature dependence of the conductivity.
(c) The band structure.

5. What sort of magnetic susceptibility graph would you expect for:
(a) MnO.
(b) Co.

6. Sketch the shape of graph you would expect for the type-2 super-conductor $YBa_2Cu_3O_7$ for each of the following:
(a) Conductivity versus temperature.
(b) Resistance versus temperature.

References

1. J. Bardeen, L. Cooper and J. R. Schrieffer, *Phys. Rev.*, 1957, **108**, 1175.
2. B. L. Chamberland, M. P. Herrero-Fernandez and T. A. Hewston, *J. Solid State Chem.*, 1985, **59** 111.
3. P. Adler, *J. Solid State Chem.*, 1994, **108**, 275.
4. S. E. Dann, D. B. Currie, M. T. Weller, M. F. Thomas and A. D. Al-Rawwas, *J. Mater. Chem.*, 1993, **12**, 1231.
5. E. Gurewitz and J. Makovsky, *Phys. Rev. B*, 1976, **14**, 2071.
6. R. Navarro, J. J. Smit, L. J. De Jongh, W. J. Craman and D. J. W. Idjo, *Physica B*, 1976, **83**, 97.

6
Non-stoichiometry

The existence of coloured specimens of normally colourless minerals, such as sodium chloride and alumina, has been known for centuries. Some sodium chloride crystals appear yellow due to the selective absorption of blue light, and sapphires/rubies are coloured forms of α-alumina. The coloured forms of these minerals are due to defects. In sodium chloride, excess sodium can lead to additional mobile electrons which can be trapped on vacant anion sites, giving the solid a yellow hue. In the gemstones, the colour is generated by substitution of a few of the aluminium cations by other trivalent cations, such as chromium, in a solid solution.

The purpose of this chapter is to explain the difference between point defects and extended defects and the effect these mechanisms have on solid state structure and properties. The principal of solid solutions will also be introduced to show how the solid state scientist can use the size and preferred geometry of cations to synthesize new materials.

Aims

By the end of this chapter you should be able to:

- Understand the differences between point and extended defects
- Describe the differences between Schottky and Frenkel defects
- Use the principal of crystallographic shear to explain the clustering of defects
- Describe the principles involving the formation of a solid solution of compounds

6.1   Thermodynamics and Imperfect Crystals

In Chapter 1, we discussed the perfect crystal. A perfect crystal has all its atoms/ions at rest, firmly rooted on their lattice sites. Even if the temperature was at absolute zero, and there were no lattice vibrations, this situation would not exist. As the temperature rises, as we have seen in Chapter 5, atoms/ions vibrate on their lattice sites and can even gain enough energy to leave their position in the lattice altogether. Therefore, the perfect arrangement certainly does not exist at any reasonable temperature.

The major reason for this is that a certain concentration of defects cause an increase in the overall free energy of the system:

$$\Delta G = \Delta H - T\Delta S$$

At first thought the introduction of vacancies in an ionic lattice looks detrimental; removing ions *decreases* the number of electrostatic interactions and the lattice enthalpy. This means ΔH becomes *more positive* and less favoured. However, this also leads to defects in the lattice which *increases the entropy* of the system. The effect is most marked when the lattice is strongly disturbed when the first defects are introduced. As the concentration of defects increases, the disruption is not as serious. The overall effect of this process is shown in Figure 6.1, where a ΔG minimum is reached for a certain number of defects. As the equation contains a $T\Delta S$ term, it follows that the defect concentration is also temperature dependent.

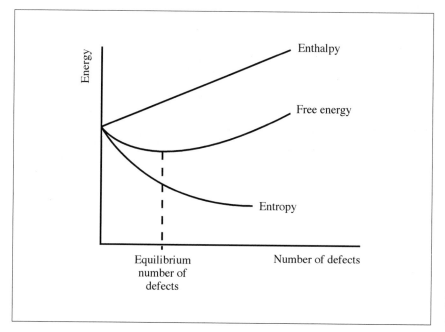

Figure 6.1 Change in energy terms with an increasing number of defects

As might be expected, compounds with high lattice energies such as highly ionic lattices have relatively few defects, as ΔH is strongly affected by the introduction of vacancies. However, there would be approximately 10,000 vacancies in one mole of sodium chloride (1.2×10^{24} ions), at room temperature, which is a significant number!

6.2   Types of Defect

Defects can be broadly classified into two types: **intrinsic** and **extrinsic**. Intrinsic defects do not effect the stoichiometry of the compound and so cannot be ascertained from the chemical formula. Often, intrinsic defects also do not affect the long range symmetry of the crystal.

Worked Problem

Q If an intrinsic defect does not affect the stoichiometry, what is an extrinsic defect?

A An extrinsic defect *does* affect the stoichiometry. This can lead to a non-stoichiometric ratio of the cations and anions or the introduction of other atoms/ions into the pure crystal.

The simplest sorts of defect are those based on **points**. Cations and anions are *removed*, or *displaced*, *randomly* in two sorts of defect.

6.2.1 Schottky Defect

Figure 6.2 shows a schematic diagram of a Schottky defect in sodium chloride. To maintain charge neutrality, equal numbers of cations and anions are removed from the crystal lattice. Although these defects have been drawn remotely separated, the defects often cluster together. The reason for this is simple electrostatics: an anion vacancy has a net positive charge as it is surrounded by six positively charged cations; conversely, a cation vacancy will have a net negative charge as it is surrounded by six negatively charged anions. These vacancies and therefore attracted towards each other by electrostatic interaction.

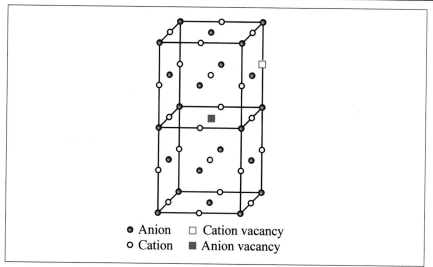

○ Anion   □ Cation vacancy
○ Cation   ■ Anion vacancy

Figure 6.2 Schottky defects consisting of pairs of cation and anion vacancies

Worked Problem

Q Is a Schottky defect intrinsic or extrinsic?

A Intrinsic. As the vacancies occur in pairs of cations and anions, charge neutrality is maintained and so is the chemical formula.

6.2.2 Frenkel Defects

The second type of stoichiometric defect involves the movement of an ion on to an interstitial site. An interstitial site is a gap in the lattice, which is not a normal lattice site. The silver halides most commonly possess this type of defect, where the silver cation moves on to the interstitial site. The cation moves into a site in the centre of a cube of cations *and* anions, as shown in Figure 6.3.

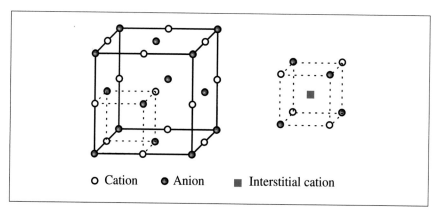

○ Cation   ● Anion   ■ Interstitial cation

Figure 6.3 Formation of cation interstitial in a Frenkel defect.

Worked Problem

Q What is the coordination number of the silver cation in the centre of the cube of cations and anions?

A The silver cation is coordinated to four cations and four anions. The Group 1 alkali metal cations, such as sodium, are small and highly charged and hence do not form this type of defect. Since the interstitial cations are in close proximity to other cations, the electrostatic repulsion of highly charged species would destabilize this structure.

Frenkel defects also occur in the fluorite structure, where the *anion* normally moves on the interstitial site, such as in CaF_2.

6.2.3 Extrinsic Defects: Non-stochiometry

Schottky and Frenkel defects do not alter the stoichiometry of the material as they are intrinsic. In non-stoichiometric materials, both types of point defect occur, but:

1. Cation and anion vacancies do not form in pairs
2. Vacancies and interstitials do not form in pairs
3. Additional dopant ions can be inserted in the structure

In general, a non-stoichiometric compound can be defined as one with *variable composition*. However, the *major structural features are maintained*. For example, Figure 6.4 shows the variation in the cubic lattice parameter (a) with oxygen content of $Fe_{1-x}O$, which crystallizes with the rock-salt structure. A smooth variation is apparent, gradually reducing as the iron content decreases.

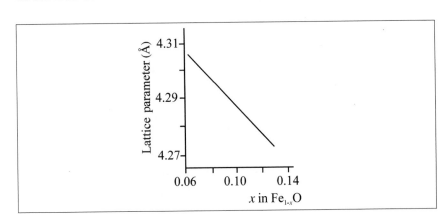

Figure 6.4 Change in lattice parameter with x for $Fe_{1-x}O$

Worked Problem

Q Why should the lattice parameter shrink in a gradual fashion?

A As iron is lost from the structure, the remaining iron in the structure is oxidized from Fe^{2+} to Fe^{3+}. The more oxidized ion is smaller owing to the contraction caused by the effect of the higher charge on the electron cloud. Also, as the defects are introduced *randomly* in the structure, the same overall structure is maintained.

Such behaviour is distinct from that of stoichiometric materials with fixed ratios such as Fe_3O_4, and Fe_2O_3, which have *different* structures.

There are four different ways in which non-stoichiometric compounds can form from a compound AB, as given in Table 6.1. These are oxidation and reduction of the metal cation, introduction of vacancies or introduction of interstitials. Most examples are in the category where the metal is oxidized and more anions are added. The formation of more bonds produces an increase in the lattice energy, which compensates for the energy required to form the defect

Table 6.1 Formation of non-stoichiometric compounds in the system AB

Oxidize A		Reduce A	
Excess anions	Cation vacancies	Excess cations	Anion vacancies
AB_{1+x}	$A_{1-x}B$	$A_{1+x}B$	$A_{1+x}B$
VO_{1+x}, $x = 0.20$	$Fe_{1-x}O$, $x = 0.14$	$Zn_{1+x}O$, $x = 10^{-5}$	ZrS_{1-x}, $x = 0.1$

Oxidation of the Metal (Insert Anion Interstitials)

Uranium(IV) oxide, UO_2, crystallizes with the fluorite structure as shown in Figure 6.5. On heating in oxygen, additional oxygen can be taken up by the lattice to form the non-stoichiometric material UO_{2+x}. There is a gradual increase in the lattice parameters as additional oxygen is added to the structure.

As discussed in Chapter 1, fluorite is a structure based on the close packing of spheres and it is not immediately obvious how the unit cell could accept extra oxygen. The biggest void in the structure is at the centre of the cube formed by the eight oxygen ions.

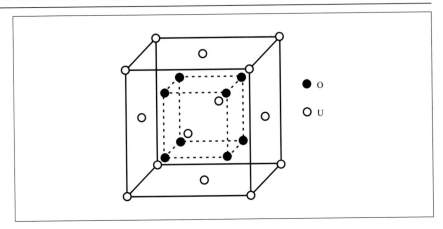

Figure 6.5 Fluorite structure of uranium(IV) oxide

Worked Problem

Q Given that the cell parameter of uranium dioxide is 5.47 Å, calculate the distance from one of the oxide ions to the centre of the cube.

A See Figure 6.6. This distance is just half the body diagonal of a cube with cell parameter $\frac{1}{2}$ the cell parameter. This gives a distance of approximately 2.37 Å.

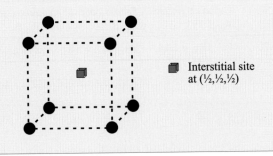

Interstitial site at ($\frac{1}{2}$,$\frac{1}{2}$,$\frac{1}{2}$)

Figure 6.6 Interstitial site in uranium(IV) oxide

The oxide ion has a radius of approximately 1.4 Å. The void at the centre of the cube would not be large enough to accept additional oxygen, as the anions would be too close together, without *movement* of the existing lattice anions. The host lattice has to 'make room' for the interstitials by creating an oxide vacancy in the host lattice. However, if the structure remained the same, this would not create enough room to put in extra oxygen, only to put in one interstitial. The stoichiometry would remain the same. The extra oxygen can only be inserted after a *rearrangement* of the fluorite lattice, where defects cluster together as shown in Figure 6.7(ii). For *every* vacancy created in the anion sublattice of the fluorite structure, *two* interstitials can now be inserted. The actual position of the interstitial anions is not precisely on $(\frac{1}{2},\frac{1}{2},\frac{1}{2})$, but is displaced relative to two specific directions. Formation of groups of defects is known as a defect cluster.

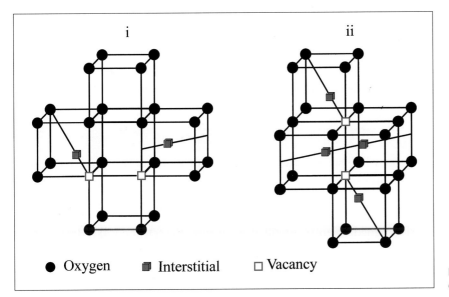

i   ii

● Oxygen   ▣ Interstitial   ☐ Vacancy

Figure 6.7 Formation of a defect cluster

Worked Problem

Q What is the formula of the compound formed from this defect cluster?

A From atom counting, the original formula would be U_4O_8:

$$U = (8 \times \tfrac{1}{8}) + (6 \times \tfrac{1}{2}) = 4 \text{ and } O = (8 \times 1) = 8$$

However, formation of the cluster results in two interstitials for every one vacancy in the fluorite lattice. Therefore host lattice = $U_4O_7 + O_2 = U_4O_9 = UO_{2.25}$.

By alteration of the host lattice and an *ordered* incorporation of extra oxygen, the compound has now become non-stoichiometric.

Oxidation of the Metal (Cation Vacancies)

$Fe_{1-x}O$ or wüstite has probably the most studied defect structure. The hypothetical compound where $x = 0$ would crystallize with a perfect version of the halite structure where iron lies in an octahedral site surrounded by oxygen atoms. Removal of some of the iron should just give iron vacancies with the remaining irons remaining in an octahedral environment.

Diffraction studies have shown that there are more iron vacancies on the octahedral sites than required by the formula of the compound. In addition, some of the iron ions appear to be in *tetrahedral* sites.

Worked Problem

Q At what position would you expect there to be vacant tetrahedral sites in this face-centred cubic lattice?

A The halite structure consists of a cubic close-packed lattice where all octahedral holes are filled. However, the tetrahedral holes which are generated by a cubic close-packed array of spheres (Chapter 1) will be empty. There are a number of available holes, for example at $(\frac{1}{4}, \frac{1}{4}, \frac{1}{4})$.

The ions in the tetrahedral sites have been shown by Mössbauer spectroscopy to have an oxidation state of +3. In a way similar to the uranium oxide structure, the distance between the closest octahedral holes and the tetrahedral hole is too short to allow both sites to be occupied simultaneously. In this case, *four vacancies* on the octahedral sites are created for *every interstitial* tetrahedral ion, as shown in Figure 6.8. This type of cluster occurs at low values of x. As x increases, larger clusters form in which there are *thirteen vacancies* and *four interstitial* ions. This is called a Koch–Cohen cluster (Figure 6.9).

Reduction of the Metal (Interstitial Cations)

Both types of reduction mechanism are more rare than the analogous oxidation mechanisms. In particular, the concentration of cation interstitials is very small. For example, x would be 0.00001 for $Zn_{1+x}O$. When zinc oxide is heated above 600 °C, the solid changes colour from white to yellow. This colour change is due to an electron transfer. On heating, the entropy term in the equation for free energy (Section 6.1) increases dramatically, resulting in oxygen vacancies. To balance the charges,

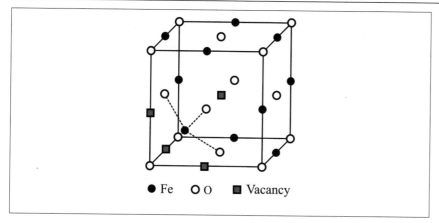

● Fe   ○ O   ■ Vacancy

Figure 6.8 Formation of vacancies in the halite structure of $Fe_{1-x}O$

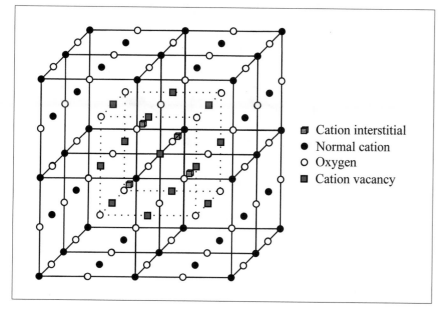

▨ Cation interstitial
● Normal cation
○ Oxygen
■ Cation vacancy

Figure 6.9 Koch–Cohen cluster

some divalent zinc ions must be reduced. These ions migrate to interstitial positions in the lattice closer to other zinc ions. The formation of the mixed valence ions in close proximity allows electron transfer between the ions. In zinc oxide, this transition is in the visible region, and hence zinc oxide appears to be yellow when hot.

Reduction of the Metal (Anion Vacancies)

VO_{1-x}, $x = 0.1$, has an oxygen-deficient lattice based on the halite structure. The distribution of vacancies is random at very high temperature, but they cluster together, along certain directions, as the temperature falls.

6.3   Anion Vacancies in More Complex Systems

Ordered anion vacancies are partly responsible for some important solid state phenomena, such as superconductivity, as well as unusual cation geometries. For example, the superconductor $YBa_2Cu_3O_7$ is built from three oxygen-deficient blocks of the perovskite (ABO_3) structure (Figure 6.10). In the parent structure, B is in an octahedral hole, coordinated to six oxygens.

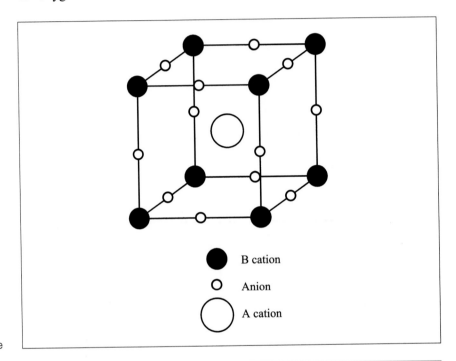

B cation

Anion

A cation

Figure 6.10 The perovskite cube

Worked Problem

Q What is the coordination number of the A cation in Figure 6.10?

A The coordination number is 12 (as A is in the close-packed layers, it must be 12; as a reminder see Chapter 1).

Ordering of the cations, and anions as shown in Figure 6.11b, and *ordered* removal of the oxygen from the BO_6 octahedra (Figure 6.11c) produces an orthorhombic material. This leads to copper–oxygen square planes and square pyramids. This arrangement is thought to be the source of its unusual electronic behaviour, where the superconducting

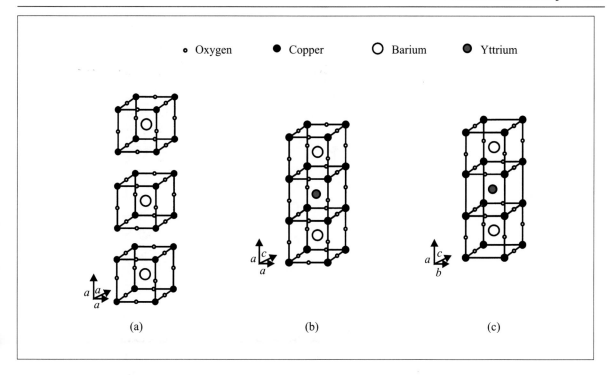

planes (CuO square pyramids) are separated by insulating one-dimensional chains (Cu–O planes). The recurring structural features of

Figure 6.11 Formation of orthorhombic $YBa_2Cu_3O_7$ by (b) cation ordering and (c) ordered removal of oxygen

Worked Problem

Q What are the coordination numbers of the two different A cations in Figure 6.11c?

A Owing to the systematic removal of oxygen from the lattice, yttrium and barium have different coordination numbers. For yttrium the coordination number is eight, but for barium it is ten.

superconducting oxides are discussed further in Chapter 7.

As the superconductivity depends largely on both the oxygen content and ordering, the preparation of a good sample of $YBa_2Cu_3O_7$ requires a low-temperature oxygen anneal at approximately 450–500 °C.

The so-called Ruddlesden–Popper phases are based on ordered arrangements of perovskite blocks and halite layers of the form $(ABO_3)_n(AO)$, where $n = 1, 2, 3$. The compound $Sr_3Fe_2O_7$, which contains iron(IV), has this structure and is shown in Figure 6.12. A graph of the change in its cell parameters with increasing oxygen content sug-

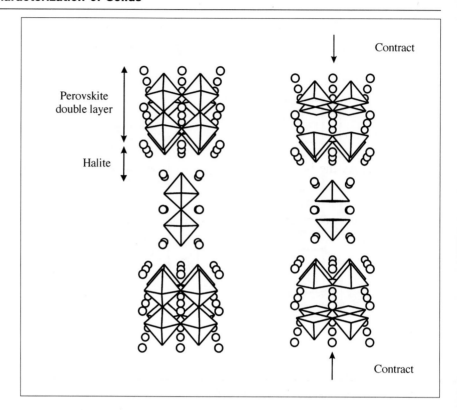

Contract

Perovskite
double layer

Halite

Contract

Figure 6.12 Contraction of the unit cell on removal of the oxygen between the perovskite blocks

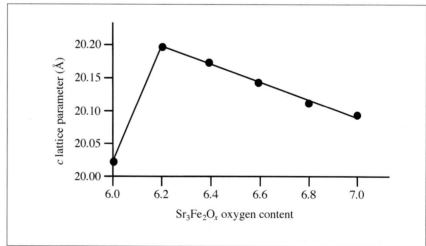

Figure 6.13 Change in the c parameter with increasing oxygen content

gests a dramatic change in the structure when a whole oxygen has been removed from the structure (Figure 6.13). This can be explained in terms of *ordered* removal of the oxygen ions from the oxygen site linking the two perovskite blocks. Initially, the lattice parameter expands as oxygen is randomly removed from the site where the iron(IV) is reduced to iron(III), where trivalent iron has a larger size due to the lower nuclear charge and hence poorer electrostatic force on the electrons. However, when all of the oxygen is removed from this site, the cell parameter shrinks dramatically. This occurs to allow better coordination of the A cation to the remaining oxygens in the square pyramids.

One problem with any material with a layered structure is that these materials often exhibit stacking faults. As the structure is the same in two dimensions, it can easily be disrupted to produce the wrong stacking sequence. As the layer sequence becomes more complex, the likelihood of stacking faults increases. Many Ruddlesden–Popper phases exist for $n = 1$, but as n increases the number of known phases decreases. This is likely to be related to the increased formation of stacking faults, which prevents isolation of the perfectly ordered phase for characterization purposes.

6.4   Colour Centres

Coloured sodium chloride crystals are due to the formation of non-stoichiometric vacancies in the anion lattice. This vacancy is capable of trapping an electron, which can then move between a number of quantized levels. These transitions occur in the visible region and generate the yellow colour. This type of vacancy in the anion sublattice of an alkali metal halide is called a Farbenzcentre or F centre. F centres can be generated by irradiation to ionize the anion, or by exposure of the lattice to excess alkali-metal cation vapour. Both procedures result in more alkali-metal cations than halide anions in the lattice.

6.5   Crystallographic Shear

Point defects and defect clusters occur randomly throughout the structure. However, it should not be surprising that these defects often cluster together. For example, the mismatch of cations and anions produced by Schottky defects produces a charge imbalance which could be eliminated by putting two defects together.

In the ReO_3 structure, the lattice is composed of corner-sharing ReO_6 octahedra. Reduction of the compound in a mixture of hydrogen and nitrogen at high temperature produces oxygen vacancies in the lattice.

Worked Problem

Q What is the formula of an infinite structure made up of corner-sharing octahedra?

A As in all infinite structures, the formula can be derived by considering the formula of a single unit cell derived from atom sharing. In a ReO_6 octahedron, all the oxygens are shared by two Re atoms. Therefore, to each Re the oxygen is only worth one-half. This gives the formula ReO_3

Every vacancy that is introduced affects two octahedra. Each rhenium atom becomes five coordinate. Dislocation of the remaining structure in the plane, perpendicular to the vacancy, produces **edge-sharing octahedra**. By performing this process, called **crystallographic shear**, each rhenium atom becomes surrounded by six oxygens (Figure 6.14).

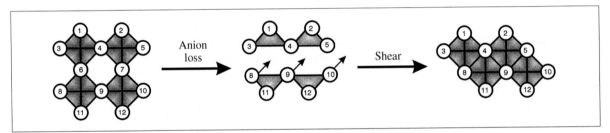

Figure 6.14 Crystallographic shear

Worked Problem

Q What is the formula of the new sheared portion of the cell?

A Consider the sheared cluster of four atoms shown in Figure 6.14. In the plane, six atoms are shared outside of the cluster with one other atom, giving $6 \times \frac{1}{2} = 3$, and four are within the cluster, counting one each. There are eight atoms above and below the cation in the plane (as it is octahedral). Hence the formula is A_4O_{15} or A_nO_{3n-1}.

Such shearing can occur randomly or in an ordered fashion. A random distribution of such shearing planes are call **Wadsley defects**. They give a *continuous* range of non-stochiometric compositions.

A family of materials exists with the formula A_nO_{3n-2} ($n = 20, 24, 25$

and 40) in which this shearing process does not occur randomly. Here sheared units are clustered together and separated by portions of AO_3 in an ordered fashion. These phases are stoichiometric ordered phases where each sheared plane is separated by a constant number of normal units. For example, $A_{24}O_{70}$ has one sheared plane separated by eleven normal units.

Worked Problem

Q Show how the stoichiometry $A_{24}O_{70}$ results from the ordered arrangement of these planes.

A Eleven normal units: $11 \times AO_3 = A_{11}O_{33}$. One sheared unit: $1 \times AO_2 = AO_2$. Added together $= A_{12}O_{35}$. Multiplying the formula by two gives $A_{24}O_{70}$.

The defects we have discussed in this chapter are largely microscopic and cannot be observed from the macroscopic structure of the materials. However, there are various sorts of macroscopic defect which can be examined using electron microscopy, and which explain certain physical characteristics. For example, metals are generally malleable and ductile but their ordered solid state structure implies that they should be rigid. Sometimes heating metals makes them more brittle in a process known as 'work hardening'. These characteristics indicate that the structures of the metals are not perfect. The malleability of metals is an indication that the structure contains defects which occur in lines and planes, allowing the atoms to slip over each other. As the temperature rises or the metal is worked (as by a blacksmith), the metal becomes harder as the defects are removed.

6.6   Solid Solutions

If two liquids are mixed, then providing they have comparable densities and polarities, a single phase is formed with no boundary between the phases. All the molecules are mixed up, and a sample of the mixture will contain the two liquids in the ratio in which they were added together. A solid solution is very similar. Atoms or ions are introduced into a parent structure and essentially disappear into it.

There are two types of solid solutions:

1.   Substitional solid solution: atoms are introduced to directly replace an atom or ion in the parent structure

2. Interstitial solid solution : atoms are introduced on to interstitial sites

Both types of mechanism can also occur simultaneously.

6.6.1 Substitional Solid Solutions

One of the simplest types of solid solution involves the substitution of ions of the same charge as those already present in the lattice. This process occurs in nature where α-alumina is doped with chromium to produce rubies. Al_2O_3 and Cr_2O_3 have the same high-temperature structure and chromium can easily substitute for aluminium in the lattice. Instead of pinpointing where the few atoms are, the composition is just written as an average, *e.g.* $(AlCr)_2O_3$, where the substitution occurs randomly throughout the lattice. This is similar to taking a sample of the liquid solution where we do not know the positions the molecules are exactly, but we know that it *must* be a mixture of what we put in.

This is the simplest sort of solid solution, where both cations have the *same charge, similar size* and *preferred coordination geometries*.

Worked Problem

Q Why might the substitution of octahedral iron(II) with copper(II) have structural implications?

A Copper(II) is a Jahn–Teller ion and normally sits in a distorted lattice site. The configuration of copper is d^9, which leads to an unpaired electron in the e_g level in an octahedral crystal field. The e_g level consists of the degenerate $d_{x^2-y^2}$ and d_{z^2} orbitals. According to the Jahn–Teller theorem, the system distorts to remove the degeneracy and lower the overall energy of the system. Hence one set of bonds (four in-plane, or two axial) become longer than the others and the copper ion has a distorted geometry.

Changes in Valency

Substitution of an ion in a lattice with another ion of different charge has implications in terms of charge balance. For example, if we substituted fluoride for oxide, then either the corresponding cations would have to be reduced or a cation vacancy would have to be introduced. More often than not, changes in oxidation state occur. Introduction of cations of different charge results either in anion vacancies (if the introduced cation has a lower charge) or anion interstitials (if the charge is higher)

Size and Coordination Geometry

Substitutional solid solutions also require the ions replacing each other to be similar in size. To a certain extent this is also a size consideration. For example, larger ions typically have higher coordination numbers.

Worked Problem

Q Which ion would you expect to have the higher coordination number, lithium or potassium?

A Potassium has an ionic radius of 1.5 Å compared with lithium at 0.9 Å. This difference is reflected in their coordination numbers, where potassium is normally six- or eight-coordinate and lithium is often found to be four-coordinate. For example, lithium oxide forms the antifluorite structure with lithium in the tetrahedral holes in a close-packed lattice of oxygen ions.

It is reasonable to assume that a difference of 30% in the radii will not allow the compound to form. A solid solution would be expected to form easily if there were very small differences in the radii, up to 10%. There is a grey region between 10% and approximately 20% where a solid solution *may* form. In this region, formation of solid solutions is a matter of trial and error.

A very important factor in the formation of solid solutions is the *temperature*. There are many solid solutions which exist at high temperature which separate into individual phases at lower temperatures. Such phases can sometimes be isolated by rapid quenching of the system, but often these phases gradually separate with time. The major reason for this is the entropy change which occurs when a solid solution is formed. The random distribution of atoms or ions over different sites in one solid has a much greater entropy than that of a mixture of the two separate phases. Owing to the temperature dependance of the entropy term, this factor dominates at high temperature. If the enthalpy term is negative, this reinforces the entropy term, and phase formation is likely to be favoured at all temperatures. However, if the enthalpy term is positive, the entropy and enthalpy are opposing, and it is only at high temperature that the entropy term wins and a solid solution is formed.

The entropy term can lead to different structures at different temperatures. For example, the alloy FeCo has a disordered structure at high temperature, and from the absences in the neutron diffraction pattern the unit cell appears to be body-centred cubic. As the temperature falls, the iron and cobalt atoms order, and a primitive unit cell is observed (Figure 6.15).

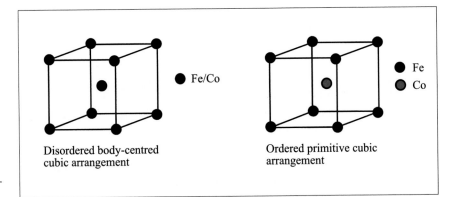

Figure 6.15 Change in the symmetry of the FeCo cubic unit cell with temperature. Shaded atoms indicate the disordered body-centred unit cell

Disordered body-centred cubic arrangement

Ordered primitive cubic arrangement

Worked Problem

Q Why would neutron diffraction have to be used for this experiment rather than X-ray diffraction?

A Iron and cobalt are neighbours in the Periodic Table, and they would be indistinguishable by X-rays. The unit cell would look disordered at all temperatures, and the pattern expected for a body-centred cube would be observed.

6.6.2 Interstitial Solid Solution

An interstitial solid solution is one in which extra atoms occupy vacant interstitial sites in a host lattice without significant movement of the host lattice ions. This process occurs with small atoms such as hydrogen and carbon in simple metal structures, as well as in more complex systems such as C_{60}. The doping of interstitial atoms on to the empty tetrahedral and octahedral sites in C_{60} is discussed in detail in Chapter 7.

Worked Problem

Q Steel is one of the most important everyday materials which depend on the interstitial solid solution of iron with carbon. Although iron exists in three polymorphs, two body-centred cubic and one face-centred cubic, only one of these forms readily absorbs carbon. By considering the structure of the body-centred cubic and face-centred cubic lattices, explain why this could be predicted.

A The face-centred cubic lattice forms octahedral and tetrahedral holes. Although it is more densely packed that the body-centred cubic version, in which the coordination numbers are 12 rather than 8, there are fewer larger holes, as shown in Figure 6.16.

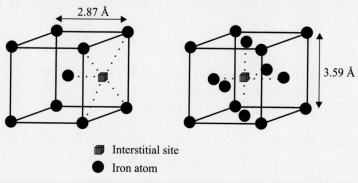

☐ Interstitial site

● Iron atom

Figure 6.16 Interstitial sites in body-centred and face-centred cubic unit cells

The distance from the nearest iron atom to the interstitial site is 1.4 Å and 1.8 Å respectively for body-centred and face-centred polymorphs. While the site in the body-centred cubic lattice is too small to accommodate the carbon without significant distortion, the site in the face-centred cubic lattice *is suitable*, and the carbon can be accommodated. Approximately one-third of the empty octahedral sites can be occupied, giving the formula Fe_3C.

Many different lanthanides and early transition metals form A_3C compounds by this type of reaction, as described for the face-centred cubic iron polymorph. The limiting factor appears to be the size of the radius of A. A critical radius of 1.35 Å seems to be required to form the face-centred cubic lattice with twelve coordinate ions at room temperature.

Summary of Key Points

1. Types of Defect
Intrinsic defects occur at points in the lattice and do not affect the stoichiometry of the material. Extrinsic defects are reflected in the chemical formula, which becomes non-stoichiometric. Extrinsic defects are caused by external influences on the lattice of the pure crystal, *e.g.* other elements or chemical treatment.

2. Defect Processes

Intrinsic defects can form by the loss of pairs of cations and anions (Schottky) or by the movement of anions or cations on to interstitial sites (Frenkel). Both mechanisms can occur simultaneously.

3. Non-stoichiometry

Non-stoichiometric defects can occur by oxidation and reduction of the cations, by introduction of interstitials, or by loss of the anions.

4. Clusters

Defects often cluster together. This can occur randomly, as in Koch–Cohen clusters, or in an ordered way by crystallographic shear to form, for example, $Mo_{10}O_{29}$.

5. Energetics

The energetics of defect formation can be exploited to make solid solutions to form new compounds.

Further Reading

A. R. West, *Basic Solid State Chemistry*, Wiley, New York, 1997.

A. K. Cheetham and P. Day, *Solid State Chemistry: Compounds*, Oxford University Press, Oxford, 1992.

M. T. Weller, *Inorganic Materials Chemistry*, Oxford University Press, Oxford, 1994.

N. N. Greenwood, *Ionic Crystals, Defects and Non-Stoichiometry*, Butterworths, London, 1968.

Problems

1. By exposing potassium chloride to an excess of potassium vapour, a compound forms which appears violet in colour. Explain the source of the colour. What sort of defect is formed?

2. 3.5 g of cerium metal reacts with 600 cm^3 of deuterium gas to form compound A, which crystallizes as a white crystalline solid. A powder neutron diffraction pattern shows that the compound crystallizes with a face-centred cubic lattice. Heating A in a stream of deuterium gas on a thermogravimetric analyser causes a weight gain corresponding to the uptake of a further 75 cm^3 of gas for the 3.5 g original sample and formation of B.

(a) Determine the formula of A and predict what structure has been formed.

(b) What sort of defect is being formed in compound B?

(c) How could the extra deuterium be incorporated into the structure?

(RMM Ce = 140; 1 mole gas occupies 24,000 cm^3 at 298 K)

3. TiO$_x$, where 0.7 < x < 1.25, crystallizes with the sodium chloride structure. By consideration of the incorporation of vacancies into the halite structure, describe how defects could be incorporated to give the formulae at the extremes of the values of x.

4. What are the differences between intrinsic and extrinsic defects. Illustrate your answer with an example of each.

5. Palladium metal crystallizes with a face-centred cubic structure. Heating palladium in hydrogen causes uptake of hydrogen into the lattice.

(a) What sites are being filled in the palladium structure?

(b) What type of solid solution is being formed?

6. What would be the formula of a compound consisting of nine normal blocks of MoO$_3$ separated by one plane of crystallographically sheared defect octahedra?

7. What kind of defects, if any, would you expect in the following crystals:

(a) KF.

(b) MoO$_{3-x}$.

(c) AgBr.

8. If a small amount of yttrium fluoride was heated over an extended period with calcium fluoride, would you expect a substitutional solid solution to form? How could the difference in charge of yttrium and calcium be accommodated?

7

Selected Topics: An Introduction to Some Important Solid State Materials

In the last 20 years, many headline-stealing discoveries have been made in the area of solid state science. For example, Nobel prizes have been awarded to two physicists, Bednorz and Müller, for their work in super-conductors and to three chemists, Curl, Kroto and Smalley, for their work on buckminsterfullerenes. Many of these discoveries lie on the boundaries between chemistry, physics and materials science. Therefore, they have required collaboration across these traditional borders to develop the new materials. This chapter selects five areas in which sig-nificant advances have been made relatively recently.

The superconductors are one of the major discoveries of the 20th cen-tury. The progress from the metals and alloys to the superconducting oxides, or so-called high-temperature superconductors, demonstrate the relationship between physical *structure* and *properties*. A second group of complex oxide materials which exhibit dramatic changes in their resis-tance with applied magnetic field, known as the giant (or colossal) mag-netoresistors, will be used to illustrate another relationship between electronic and physical properties.

Finally, three different sorts of compound which can insert and exchange cations and ions in their structures will be discussed. Of these, the zeolites have been developed from minerals, and are used as ion exchangers, catalysts and sorbents. The use of their *framework* struc-tures for shape-selective catalysis will be discussed. The unique proper-ties of both layered and three-dimensional compounds which can accept *extra* ions, such as graphite and C_{60}, will also be examined.

7.1 Superconductivity

As discussed in Chapter 5, the conductivity of some compounds changes abruptly at a particular temperature. Below this critical temperature, they conduct with no appreciable resistance. These compounds are called superconductors. In the superconducting state, the compound offers no resistance to the flow of conducting species. A current started in a loop will therefore continue to flow indefinitely, without significant loss. This property, and others particularly associated with the superconducting state, could theoretically allow superconducting materials to be used for many applications, such as:

1. By replacing copper circuitry with superconductor circuitry, no energy would be wasted.
2. Infinite current could be passed in an electromagnet, creating much stronger magnetic fields.
3. Using the Meissner effect, trains could levitate above magnetic rails and require only minimal electricity to move.

7.1.1 A Brief History

While Onnes[1] was experimenting on the liquefaction of helium in 1911, he found the resistance of mercury dropped dramatically from 0.08 Ω at 4.2 K to less than 3×10^{-6} Ω at 4 K over a temperature interval of 0.01 K (Figure 7.1). He named this phenomenon superconductivity. This behaviour is the most striking feature of superconducting materials, in which below a critical temperature, T_c, the electrical resistance *suddenly* drops to effectively zero.

Over the next 60 years, superconductivity was observed in many alloys, metallic elements and intermetallic compounds with T_c values in

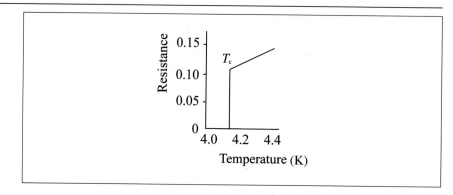

Figure 7.1 Change in resistance of mercury with temperature

the range 1–18 K. However, a limit seemed to have been reached with the intermetallic compound Nb_3Ge, with a critical temperature of 23.3 K (Figure 7.2).

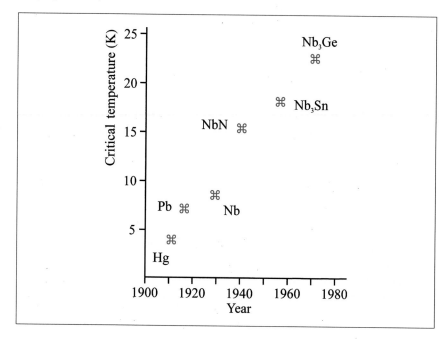

Figure 7.2 Critical temperatures of some elements and intermetallics

This stood as a maximum until 1986, when Bednorz and Müller[2] reported 'the possibility' of superconductivity in the Ba–La–Cu–O system. Soon, confirmation of the superconducting phase $La_{2-x}Ba_xCuO_4$ was heralded by many other laboratories worldwide, and a record T_c of 35 K was achieved with $x = 0.4$.

A few months later the superconductor $YBa_2Cu_3O_7$ was discovered,[3] with a T_c of 93 K. The discovery of this compound was the birth of high-temperature superconductivity (superconductivity above the boiling point of nitrogen). Frenzied activity spanning many countries and multimillions of pounds then led to a variety of high-temperature superconductors with higher and higher T_c values (Figure 7.3).

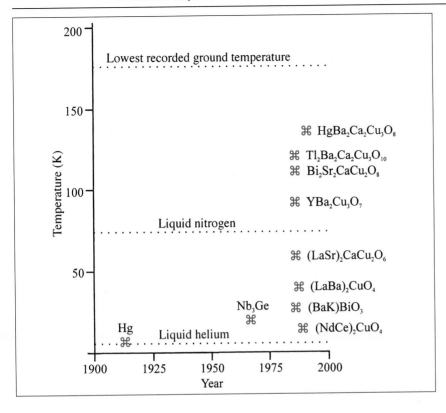

Figure 7.3 Evolution of T_c with time

Although the rapid increases in T_c were not sustained, the understanding of the chemical features and formation of high-T_c materials has made major advances in the last decade.

7.1.2 Properties of Superconductors

Critical Temperature

In a normal metallic conductor the electrons move through the lattice, giving the conducting current. The flow of electrons is interrupted by defects and vibrations of the atoms on the lattice points. As the temperature increases the vibrations increase and the resistance rises. In the superconducting state, below the critical temperature, the conducting current is thought to move *with* the lattice vibrations in a concerted process, so that there is no resistance. This is the Bardeen, Cooper and Schieffer (BCS) theory,[4] which was discussed in detail in Chapter 5. This theory was developed before the advent of high-temperature superconductors. It explains type 1 superconductivity well; type 1 superconductors are either superconducting or not superconducting at a particular temperature and applied field. However, it fails to explain the gradual loss of superconductivity observed for the type 2 materials.

Magnetism: The Meissner Effect[5]

Magnetic flux is excluded from the sample by the setting up of surface currents which exactly oppose the magnetic field. This results in a strong repulsive force, similar to the poles of a magnet. The Meissner effect is powerful enough to deflect a permanent magnet, and can even levitate a magnet above the superconductor surface.

7.1.3 Elements and Alloys

Table 7.1 indicates some elements of the Periodic Table which have been shown to have a superconducting transition under normal conditions of temperature and pressure. Other elements exhibit superconductivity under exceptional conditions, *e.g.* under pressure (Si, Y), or when prepared as thin films (Li, Cr) or irradiated by α-particles (Pd). T_c for the elements is generally below 10 K (maximum Nb, T_c = 9.25 K). Small increases in these critical temperatures can be achieved by using high pressure to force the atoms closer together.

Table 7.1 Selected elements and their critical temperatures (K)

Element	T_c	Element	T_c
Nb	9.25	Al	1.18
Pb	7.2	Os	0.66
Hg	4.15	U	0.20
In	3.40	Rh	0.0003
Re	2.57		

The first indicator as to what *does not* make a superconductor is found in the transition elements; those elements with incomplete d or f shells which order ferromagnetically are not superconductors at any temperature. In fact, until very recently (1997), magnetism and superconductivity were thought to be mutually exclusive. However, two intermetallics consisting of ruthenium/copper and indium/gold have recently been discovered which are both magnetic *and* superconducting.

Solid solutions between elements in the Periodic Table which are superconductors are very important materials. Some of the most common superconducting materials are in this class, in particular those containing niobium. For example, NbTi and NbZr are fabricated to form superconducting wires for use in coils.

Intermetallic compounds composed of metal/metal or metal/non-metal form another group of superconducting compounds. The most important of these crystallize with the β-tungsten structure. The stoichiometric formula is A_3B, where the A element is nearly always a transition metal of Groups 4, 5 or 6. The B atoms lie in a body-centred cubic arrangement, and two equidistant A atoms sit in the middle of each face. They form three non-intersecting continuous chains (Figure 7.4). The overall lattice is primitive as the A atoms are not related by $1/2, 1/2, 1/2$ translation (see Chapter 1). To date, Nb_3Ge has the highest T_c of these materials at 23.3 K.[6]

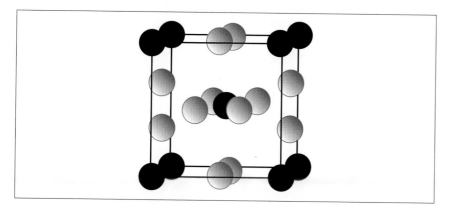

Figure 7.4 A_3B structure

7.1.4 Superconducting Compounds

There are a vast number of different sorts of superconducting compound. Examples of these are summarized in Table 7.2. They include organic polymers and intercalation compounds (such as Rb_3C_{60}), as well as ceramic oxides and sulfides. In all these compounds the critical temperature is still below the boiling point of liquid nitrogen. Hence, these materials would be very expensive to use in everyday life (as a litre of liquid helium costs 20 times as much as liquid nitrogen).

Table 7.2 Examples of non-cuprate-based superconductors

Structure	Symmetry	Example compound	T_c (max) (K)
NaCl-type	Cubic	NbN, SrN, GeTe, PdH	17.3
WO_3 (tungsten bronze)	Hexagonal	Rb_xWO_3	7
Spinel	Cubic	$Li_xTi_{3-x}O_4$	13.7 (x = 1)
Perovskite	Cubic	$Ba_{1-x}K_xBiO_3$	30
Chevrel phase	Tetragonal	$PbMo_6S_8$	15.2
C_{60}	Cubic	$(RbCs)_3C_{60}$	33
Complex	–	$(BEDO\text{-}TTF^*)_2ReO_4$	13

* Bis(ethylenedioxy)tetrathiafulvalene.

One important point to note is that, despite the great diversity of these compounds, their structures are relatively simple (except for the organic materials). This meant that in the hunt for new superconductors, many new materials were rapidly synthesized in these families of compounds.

The real breakthrough came in 1986, with the discovery of high-temperature superconductivity in materials containing copper and oxygen (hereafter called **cuprate** materials). Since then, a variety of materials have been discovered which are, at first sight, bewildering in their complexity. However, all these structures are actually relatively simple. Most are built from **perovskite** (ABO_3) structural units, as shown in Figure 7.5, together with, in some cases, **rock-salt-like** layers (Figure 7.6). The perovskite structure is based on closed-packed layers of alternating AO and O, with the octahedral holes filled with the smaller B cations. Combining this structure with rock-salt layers produces compounds with very long c-axes (in comparison with a and b) owing to the stacking of different layers.

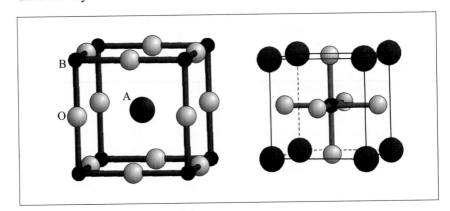

Figure 7.5 The perovskite structure

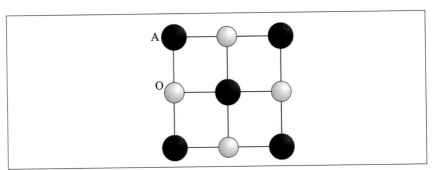

Figure 7.6 Rock-salt layer

Unfortunately, many of these materials contain unusual and toxic elements such as mercury, thallium and lead. The current record is for a mercury-based compound with a T_c of 133 K at atmospheric pressure, increasing to 160 K under pressure. Some superconducting cuprates are summarized in Table 7.3.

Table 7.3 Examples of high-temperature superconducting cuprates

Compound type	Structural base	Highest T_c
$La_{2-x}Ba_xCuO_4$	Perovskite and halite	$x = 0.4$, 35 K
Yttrium based	Perovskite	$YBa_2Cu_3O_7$, 93 K
Bismuth based	Perovskite and halite	$Bi_2Sr_{2-x}Ca_{2+x}Cu_3O_{10}$, 110 K
Thallium based	Perovskite and halite	$Tl_2Ba_2Ca_2Cu_3O_{10}$, 125 K
Mercury based	Perovskite and halite	$HgBa_2Ca_2Cu_3O_8$, 160 K

Nomenclature

Owing to both the complexity of the superconducting structures and the number of closely related materials containing essentially the same elements, a way of simplifying the names was quickly found. This uses the number of moles of element in the formula, preceded by a qualifying element. For example, the thallium family would start 'thallium'. All the other elements (excluding oxygen) are then given numbers. $Bi_2Sr_{2-x}Ca_{2+x}Cu_3O_{10}$ would therefore be called bismuth 2223. If one of the alkaline earth elements is missing, a 0 would be included.

Worked Problem

Q How could the phases $YBa_2Cu_4O_8$ and $Bi_2Sr_2CuO_6$ be given a shorter name?

A $YBa_2Cu_4O_8$ is simple and would just be called yttrium 124 from the cations in the formula. $Bi_2Sr_2CuO_6$ is slightly more difficult. The calcium normally present in the formula of the bismuth materials is absent. This phase would therefore be called bismuth 2201 where the 0 corresponds to the missing calcium.

Structural Aspects

Certain structural, magnetic and electrical features have been recognized as recurrent in the cuprate-based materials:

1. Electronically active copper–oxygen sheets, which are thought of as the centroid of superconductivity.

alternating with:

2 Insulating block layers, which are as important as they act as reservoirs of positive or negative charge. They control the electronic

make-up of the material, adding to or removing electrons from the copper–oxygen planes.

3. The interspacing of the layers is generally 3.6 Å.

A schematic diagram of a superconducting oxide is shown in Figure 7.7, together with some actual superconductors, indicating the important structural features.

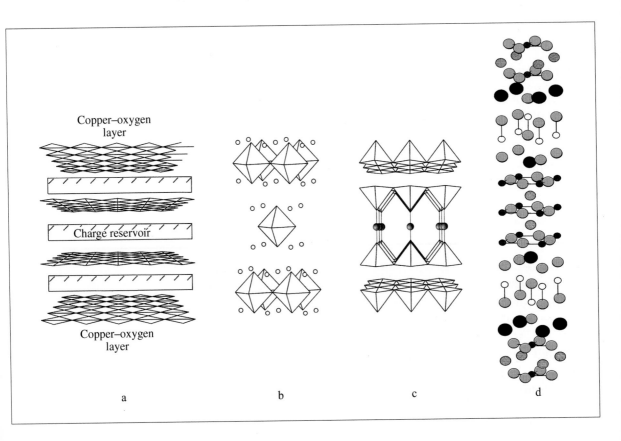

Copper–oxygen
layer

Charge reservoir

Copper–oxygen
layer

a   b   c   d

Figure 7.7 Schematic diagram of a superconductor (a) and some cuprate materials:
(b) $La_{2-x}Ba_xCuO_4$; (c) $YBa_2Cu_3O_7$;
(d) $Bi_2Sr_2Ca_2Cu_3O_{10}$

7.1.5 Charge Carriers

The high-temperature superconductors can be classified by the two different types of charge carrier: positive and negative. In the former the charge carriers are positive holes, but in the latter they are electrons. Positive charge carriers are the most common.

Positive-type: Holes

The mechanism of hole-based charge carriers is simple. If you take a parent compound which is non-metallic and non-superconducting and introduce holes in the copper–oxygen sheets in a controlled manner (*e.g.*

by changing the chemistry of the insulating layer), then you create a hole. If these holes can move, a conductor is produced.

For example, La_2CuO_4 is an antiferromagnetic insulator. Holes can be introduced in the copper sheets by doping in divalent barium, or strontium, for trivalent lanthanum or by introducing oxygen interstitials. For each divalent ion we have either one Cu^{3+} or one O^-. Therefore, for $La_{2-x}Sr_xCuO_4$ the hole concentration is directly proportional to x. At $x = 0$, the compound is an antiferromagnetic insulator. As x increases, the material becomes a non-magnetic insulator and at $x = 0.4$ a superconductor. This generates a *compositionally controlled* insulator–metal transition.

Hole-based superconductors have an average copper oxidation state of approximately 2.2. The holes are normally introduced by oxygen vacancies in the apical oxygens of the copper–oxygen layers.

Worked Problem

Q Calculate the average copper oxidation state in the compounds $YBa_2Cu_3O_{6.5}$ and $YBa_2Cu_3O_7$ and hence predict whether they could be hole-based superconductors.

A Taking the oxidation states of yttrium, barium and oxygen as 3, 2 and –2, respectively, the average copper oxidation states are 2 and 2.33 for the two compounds. The first compound has an integral copper oxidation state and would not be expected to be superconducting. In fact, as oxygen is lost from $YBa_2Cu_3O_7$ the critical temperature falls until the compound becomes non-superconducting when the oxygen content in the formula drops to O_6.

Negative-type: Electrons

Electron superconductors, such as $Nd_{2-x}Ce_xCuO_4$, do not have oxygen vacancies and the formal oxidation state of copper in these compounds is less than 2. Conduction is by the electron current.

7.1.6 Superconducting Oxide Tape

Unlike the alloys and elements discussed previously, it is not immediately clear how a superconducting oxide could be made into a wire. Superconducting oxides are *powders*. The problem is overcome by making tapes, rather than wires, and using an exterior coating of silver. Silver is used as it is practically the only material which does not react with the superconductor during heating. The powder is poured into a silver tube and then rolled down, in a similar way to rolling dough, to form a thinner tube. The tube is then heated and flattened to form a tape. Many tapes are then bundled together to form a superconducting wire. A direct comparison of a superconducting wire and a conducting wire of the same thickness is startling. The superconducting wire can carry 1000 times the current!

7.1.7 Uses of Superconductors

There have been many projected uses for superconducting compounds, including a.c. and d.c. cables, generators, motors, transformers, energy storage, wires, computers and magnetically levitating trains to name just a few. Some detailed examples are given below.

Magnetic Resonance Imagers

Superconductors have revolutionized the field of biomagnetism. Magnetic resonance imaging (MRI) is a non-invasive way of examining soft tissue. Using a strong magnetic field, the hydrogen atoms in water and fat molecules accept magnetic energy, which can be detected when the energy is released. The technique was developed in the late 1940s, but conventional magnets created a field approximately one quarter of the strength of those possible with superconducting magnets. This meant the early MRI machines were closed devices, where the patient was enclosed for long periods inside the magnet. Owing to the higher field of the superconducting magnets, the devices can be more open and data collected in a fraction of the time, with increased patient comfort.

Superconducting Magnetic Energy Storage (SMES)

The purpose of these storage systems is to damp power fluctuations in the power grid to prevent computer crashes and equipment failure. Energy is stored in powerful electromagnets and can be retrieved whenever there is a need to stabilize line voltage during unanticipated disturbance in the power grid.

Electronic Filters

The discrimination provided by normal electronic filters is limited. Addition of each filter to reduce noise diminishes the overall signal. As superconductors have zero resistance, superconducting filters do not diminish the overall signal, and hence many more filter stages can be added. One possible use for this technology would be in cellular telephone systems, where it would allow more remote areas to be accessed.

MagLev Trains

Magnetic levitation is an application which could theoretically be used for people transport. Transportation vehicles, such as trains, would be made to float on strong superconducting magnets, eliminating friction between the train and its tracks. This would mean extremely high speeds could be achieved. This was realized in 1990, with a nationally funded project in Japan to develop just such a train. The Yamanashi test line opened in April 1997. On 14 April 1999 a speed of 340 miles per hour was achieved with the MLX01.

7.2   Magnetoresistors

The change in the electrical resistance of a material in the presence of a magnetic field is called its magnetoresistance (MR). For certain groups of materials, termed giant (GMR) or even colossal magnetoresistors (CMR), the difference in resistance is enormous (10^6 Ω). The term 'giant' or 'colossal' just refers to the different orders of magnitude of the change in resistance. Until recently, these effects were thought to be limited to intermetallics and doped semiconductors.

   The discovery of giant magnetoresistance in the perovskite-based material La–Sr–Mn–O prompted intense research similar to that for new high-temperature superconductors. Magnetoresistance has since been found in many other oxide and sulfide materials, such as $Tl_2Mn_2O_7$[7] and $FeCr_2S_4$.[8] However, it is the manganese perovskites which have received the most attention. Possible uses for magnetoresistors include actuators, superior capacity hard-disks and sensors.

7.2.1 The Zener Mechanism[9]

The doped perovskite structure is shown in Figure 7.8. It consists of an infinite network of manganese–oxygen octahedra. According to the stoichiometry, and for charge balance in the formula $La_{0.7}A_{0.3}MnO_3$, A = Sr and Ba, the material must contain Mn^{3+} and Mn^{4+}. If we consider the electronic configuration of these ions in a octahedral ligand field (Figure 7.9), it is immediately apparent that trivalent manganese contains four d-electrons. This would give manganese the typical electronic structure expected to undergo Jahn–Teller distortion, where manganese has one electron in the degenerate e_g set (upper level on the figure). According to the Jahn–Teller theorem, the site will undergo distortion to remove the degeneracy. This is where the source of magnetoresistance

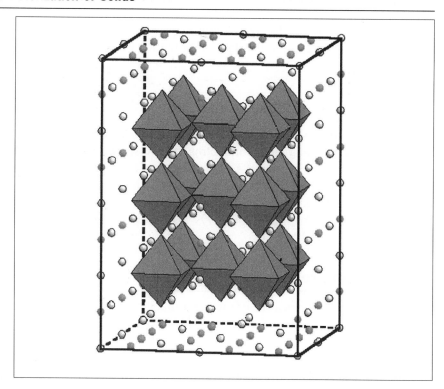

Figure 7.8 (LaSr)MnO$_3$ structure containing vertex-linked MnO$_6$ octahedra

is thought to lie. When an electron hops from the symmetrical octahedral arrangement of Mn^{4+} to the asymmetric, Jahn–Teller distorted, arrangement of Mn^{3+}, the new site has the wrong symmetry (see Figure 7.9). Therefore, local distortion must occur around the manganese centres when the electron moves. This is shown schematically in Figure 7.10 and is called the Zener mechanism.

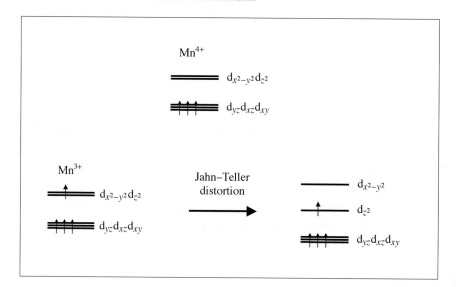

Figure 7.9 Electronic configuration and Jahn–Teller distortion of Mn^{3+}

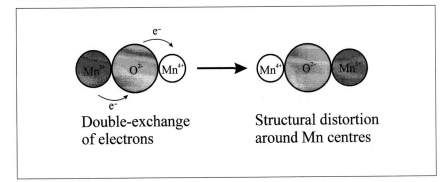

Double-exchange
of electrons

Structural distortion
around Mn centres

Figure 7.10 Schematic diagram
of the Zener mechanism

However, the mobile or localized behaviour of the electrons changes
so dramatically with only minor changes in the stoichiometry (Figure
7.11) that the Zener mechanism is thought to be too simple. It also fails
to explain the magnetoresistance in compounds in which there are no
Jahn–Teller distorted ions.

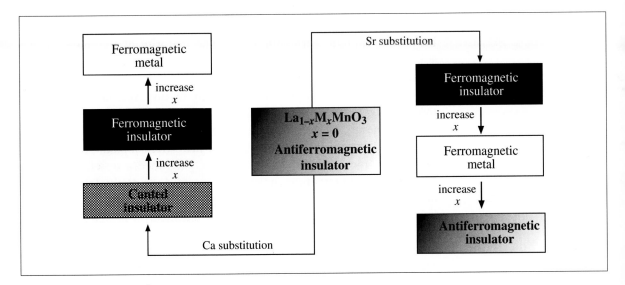

Figure 7.11 Changes in electronic behaviour with composition for
the alkali-earth doped lanthanum
manganate perovskites

Worked Problem

Q What is the oxidation state of Mn in $Tl_2Mn_2O_7$?

A With thallium trivalent, manganese must be present as Mn^{4+}.
This is not a Jahn–Teller ion, since there are three d-electrons, one
each for the t_{2g} set of d-orbitals, and so the Zener mechanism would
not explain the magnetoresistance of this compound.

7.3 Zeolites

The zeolites are a group of compounds, some of which occur naturally, which are named after their ability to evolve water when heated. The name 'zeolite' comes from the Greek *zeo* to boil and *lith* stone. These materials form an extensive group. Both natural and synthetic zeolites are used for ion exchange, absorption and catalytic purposes. Although Cronstedt discovered the naturally occurring minerals as far back as 1756,[10] and some were used in the early 1800s in Japan as air purifiers, it was not until this century that the real potential of zeolites was recognized. Much of the development of zeolites is credited to Barrer, whose pioneering work began in the late 1930s.[11]

Zeolites are compounds with *low molar masses*. Their structure consists of frameworks that form cavities/channels, which may incorporate a range of small inorganic and organic species. The frameworks are constructed from linked tetrahedra. Many elements which form TO_4 groups, where T is a tetrahedral atom, can be the building block in zeolites. The most common groups are AlO_4, SiO_4, BO_4 and PO_4, but T can also be beryllium, gallium, germanium, *etc.* The majority of the materials are solely based on silicon and aluminium. These also form the largest group of naturally occurring materials, which are the true zeolites. Materials containing elements other than aluminium and silicon are correctly termed zeotypes, reflecting the similar structure.

The general formula for zeolites may be obtained by starting from pure silica, in which all the tetrahedra are vertex linked. Replacement of some of the SiO_4 tetrahedra with AlO_4 tetrahedra upsets the charge balance, so countercations have to be inserted inside the framework to maintain charge neutrality; for example, if you replace an 'SiO_2' unit (where each of the four oxygens in the tetrahedron is shared with one other unit, so giving $Si + [4 \times \frac{1}{2}]O = SiO_2$) with an '$AlO_2^-$' unit, then a monovalent cation has to be added to maintain the charge. Mathematically, this gives the following formula:

$$[M^{n+}]_{x/n} [AlO_2]_x [SiO_2]_{1-x}$$

Typically, when the cations are introduced they are hydrated. Variable amounts of water are represented by modification of this formula to also include an '$.mH_2O$' term. A schematic diagram of a zeolite structure with cavities and cations is shown in Figure 7.12.

By dehydrating the zeolite, the cations are forced closer to the framework in order to achieve better coordination to the framework oxygen atoms. This means that dehydrated zeolites can often absorb small molecules other than water in their pores. Obviously, the larger the cavities in the zeolite, the larger the molecules it can absorb.

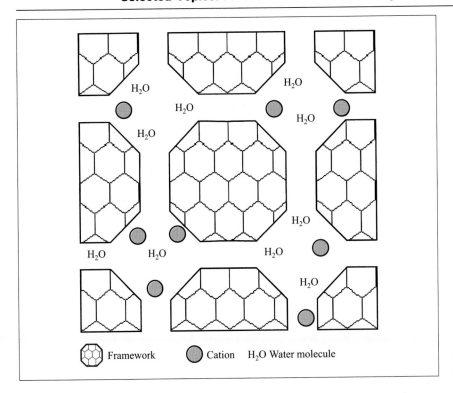

Figure 7.12 Schematic diagram of a zeolite

7.3.1 Ring Size and Structure

The tetrahedral units can link together to form a number of different sized rings, the most common of which are shown in Figure 7.13. Each black line represents a T–O–T link. These rings are then linked together to form three-dimensional structures. As there are a variety of ways in which the rings can be linked, and the T–O–T links are flexible, many structures and many different cations can be incorporated.

The most common zeolite structure, which is made up from four- and six-membered rings, is the sodalite cage, which is shown in Figure 7.14. This unit is often called the β-cage, and is itself a building block in the

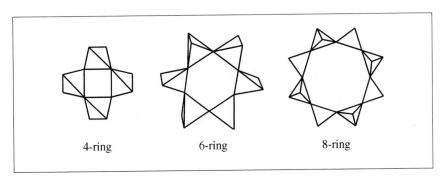

4-ring   6-ring   8-ring

Figure 7.13 Rings constructed from tetrahedral units

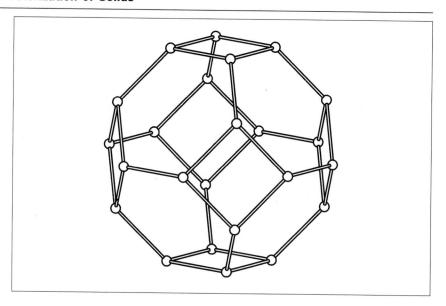

Figure 7.14 Sodalite cage

structures of other zeolites. Some larger zeolites contain the sodalite cage as a *building block*, for example zeolite A, shown in Figure 7.15. The sodalite cage is separated by other rings in these structures; in zeolite A these rings are four membered, and in faujasite they are six membered. Other examples of the linking units, or *secondary building blocks*, are shown in Figure 7.16. The effect of these linking units is to increase the size of the cavities and pores.

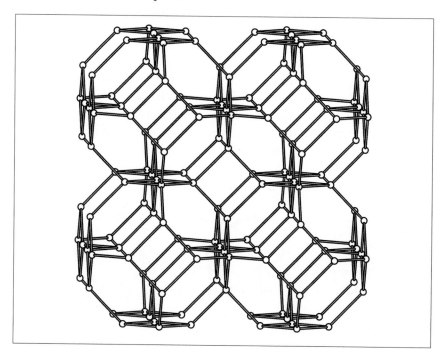

Figure 7.15 Zeolite A containing the sodalite cage as a building block

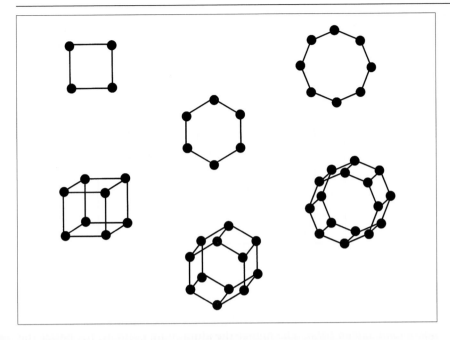

Figure 7.16 Secondary building blocks

Although many zeolite structures are cubic and their channels are the same in all three lattice directions, materials such as ZSM-5 have more complex structures. Here the channels, which are parallel to the axes, have different sizes. Large channels parallel to the b-axis intersect smaller channels in the other two directions. ZSM-5 is unusual in that the synthetic form of the structure was synthesized some 20 years before the natural form was discovered in the mineral mutinaite.

7.3.2 Lowenstein's Rule[12]

Lowenstein's rule states that the formation of Al–O–Al links is thermodynamically unfavoured. This means that the majority of zeolites (which are prepared at low temperatures using hydrothermal methods) have *ordered* frameworks, with no Al–O–Al links, and the highest Si:Al ratio is 1:1.

7.3.3 The Properties of Zeolites

Adsorption

The open framework structure of the zeolites allows small molecules to be adsorbed into their structures. The size and shape of the molecules adsorbed depends on the structure of the zeolite, and hence the geometry of its pores. For example, zeolite A readily adsorbs water but ethanol is excluded. Figure 7.17 demonstrates the effect of cavity size on the molecules which can be adsorbed.

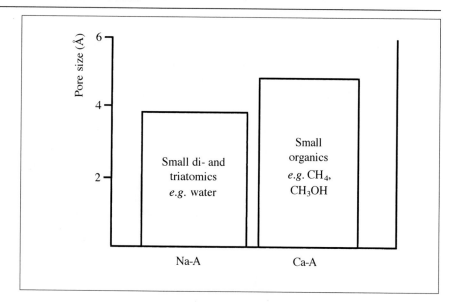

Figure 7.17 Effect of cavity size on adsorbed species for zeolite A.

The *ability* of a zeolite to adsorb water is strongly correlated to its *aluminium:silicon ratio*. The higher the aluminium content, the better the adsorption capability. Zeolite A, with an aluminium ratio of 1:1, is used widely as a drying agent for drying both gases and solvents. The drying capacity falls off with time as the zeolite becomes more hydrated, but can be restored by heating to drive off the adsorbed water.

Zeolites which are almost totally silicon based have few cations in the cavities and are hydrophobic. These materials readily adsorb non-polar solvents such as benzene.

Ion Exchange

Undoubtedly the most widescale use of zeolites is as ion-exchange reagents. Approximately 60% of the zeolite market is used for ion exchange (Figure 7.18).

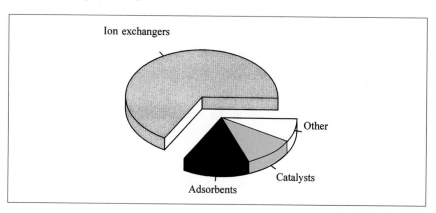

Figure 7.18 Zeolite usage

For example, naturally occurring clinoptilolite is used by BNFL to remove radioactive caesium and strontium from spent pile effluent. However, the synthetic zeolites, A and P, have had the greatest impact on the home. Sodium zeolite A rapidly exchanges its sodium atoms for the divalent hard water ions calcium and magnesium at room temperature. This discovery led to the replacement of the polyphosphates previously used in washing powder, which were damaging to the environment, by zeolite A about 20 years ago. More recently, Unilever introduced MAP (maximum aluminium zeolite P) as a replacement for zeolite A. MAP has a lower water content, and is lighter in the packet:

$$\text{Na - Zeolite A} \ + \ \tfrac{1}{2}\text{Ca}^{2+} \ \rightarrow \ \text{Ca}_{0.5}\text{- Zeolite A} \ + \ \text{Na}^+$$

Catalysis

The acidic derivatives of zeolites, H-zeolites, are excellent catalysts and are widely used in industry. The acidic form is easily obtained by direct exchange with weak acids. However, since many zeolites are broken down by acids, exchange with ammonium salts, followed by removal of ammonia at 500 °C, is often used to generate the H-zeolite.

The nature of this catalytically active zeolite can be either Brønsted (proton donor) or Lewis (electron-pair acceptor) acidic, with protons attached to the framework tetrahedra or not (Figure 7.19).

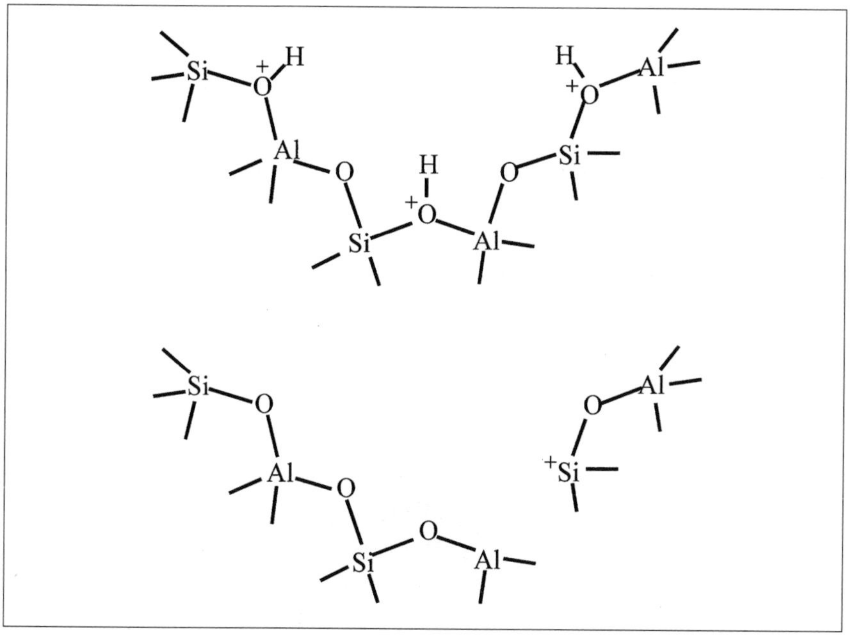

Figure 7.19 Lewis and Brønsted acid sites

Once generated, the acidic zeolites can absorb molecules into the cavities and behave like strong acids. Major reaction types are *rearrangement* and *dehydration*.

However, a very specific sort of catalytic behaviour peculiar to clays and zeolites is *shape-selective* catalysis. Essentially, owing to the fixed nature of the framework channels and cavities, only molecules of the right geometry can pass through. As only certain molecules will have the right geometry to pass into the zeolite, only these molecules will reach the active sites. The selectivity can occur either in reactant or product, or even the transition state.

For example, in the isomerization of dimethylbenzene the 1,4-isomer is the only isomer which can pass through the channels unhindered. The other molecules get stuck and interact with the active sites. Hydrogenation of the benzene ring allows the methyl groups to migrate round the ring. This means that gradually the mixture of isomers (racemate) is converted to the single isomer 1,4-dimethylbenzene.

7.4   Intercalation Reactions

Intercalation reactions are reactions of solids in which a guest molecule or ion is inserted into a solid lattice *without* major rearrangement of the solid state structure. This process is unusual for solid state chemistry as it does not require excessive bond breaking, but the host structure *must* have special *chemical* and *structural* features. In particular, intercalation reactions require the host lattice to have:

1. A strong covalent network of atoms which remain unchanged during the reaction.
2. Vacant sites which are interconnected, and of suitable size to allow diffusion of the guest species into the host lattice.

As a consequence, some of these reactions occur at room temperature, but higher temperatures can be used as long as sufficient energy is not imparted to destroy the covalent lattice.

Despite these restrictions, many families of compounds undergo intercalation reactions including: chain structures, layered lattices and three-dimensional connected frameworks with tunnels or channels.

7.4.1 Layered Structures

The layered structures form the largest group of compounds which undergo intercalation reactions. Some of the commonly observed examples are given in Table 7.4. The special feature of these materials is that *very large* molecules can be incorporated by *free adjustment* of the interlayer separation. All these materials have very *strong intralayer*

Table 7.4 Intercalation compounds

Neutral layers	
Graphite	
$Ni(CN)_2$	
MX_2	M = Ti, Zr, Hf, V, Nb, Ta, Mo, W; X = S, Se, Te
MPX_3	M = Mg, V, Mn, Fe, Co, Ni, Zn, Cd, In; X = S, Se
MoO_3, V_2O_5	
$MOXO_4$	M = V, Nb, Ta, Mo; X = P, As
MOX	M = Ti, V, Cr, Fe; X = Cl, Br

Negatively charged layers	
AMX_2	A = Group 11; M = Ti, V, Cr, Mn, Fe, Co, Ni; X = O, S

Clays and layered silicates	
Titanates	e.g. $K_2Ti_4O_9$
Niobates	e.g. $K[Ca_2Na_{n-3}Nb_nO_{3n+1}]$; $3 < n < 7$
$M(HPO_4)_2$	M = Ti, Zr, Hf, Ce, Sn

Positively charged layers	
Hydrotalcites	$LiAl_2(OH)_6OH.2H_2O$, $Zn_2Cr(OH)_6Cl.2H_2O$

bonding (in the layers) but *weak interlayer* interactions (between the layers).

The layered compounds fall into two categories:

1. *Neutral compounds*: the interactions between the layer are primarily of the van der Waals type, and the interlayer space is an array of empty lattice sites.
2. *Charged compounds*: the layers are held together by weak electrostatic forces, and the interlayer sites are partially or completely filled with ions and/or solvent molecules.

Two possible types of reaction occur on intercalation:

1. Ion exchange or molecule exchange reactions, where the charge on the framework layer remains unchanged, *e.g.* sheet silicates, clays.
2. Intercalation into structures such as transition metal structures or graphite, where the layer is *oxidized* or *reduced*.

Type 1: Muscovite

The clay muscovite undergoes simple ion exchange reactions by substitution of the ions between the layers, as shown schematically in Figure 7.20. It consists of lamellar-like layers with water molecules and cations between the strata. These cations can be easily exchanged.

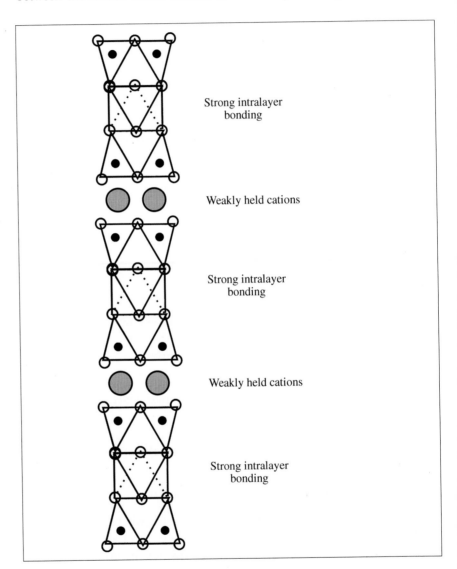

Strong intralayer bonding

Weakly held cations

Strong intralayer bonding

Weakly held cations

Strong intralayer bonding

Figure 7.20 Schematic diagram of the muscovite mica

Type 2: TaS$_2$

Many transition metal sulfides have layer-type structures which consist of layers of edge-sharing MS$_6$ octahedra held together by weak van der Waals forces, as shown in Figure 7.21.

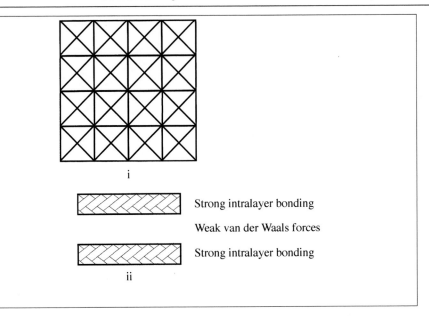

Strong intralayer bonding

Weak van der Waals forces

Strong intralayer bonding

Figure 7.21 (i) Layer of edge-sharing MS_6 octahedra, (ii) perpendicular view of the slab structure

Worked Problem

Q Show that the formula for an infinite layered lattice of edge-sharing octahedra is AB_2.

A Each of the four B atoms in the plane is shared between four octahedra. To each octahedron they are worth one quarter. At the apex of each octahedron is another B which is shared between two octahedra. For each unit, $A = 1$, $B = (4 \times \frac{1}{4}) + (2 \times \frac{1}{2}) = 2$. Hence AB_2.

These layered materials can insert a variety of other materials between the layers, including:

1. Alkali metals, by direct reaction with the metal and sulfide in a sealed tube, by electrochemical reaction in a non-aqueous solvent or by reaction in liquid ammonia.
2. Transition metal salts, *e.g.* $FeCl_3$.
3. Amines, *e.g.* pyridine or $C_{2n}H_{2n+1}NH_2$ ($n = 0–9$). When $n = 9$ the expansion of the lattice parameter is in excess of 50 Å. The chains lie perpendicular to the layers, giving a structure reminiscent of a concertina (Figure 7.22).
4. Organometallics, *e.g.* cobaltocene, which is an unstable 19-electron system and reacts at room temperature with TaS_2 by donating the extra electron to the layers.

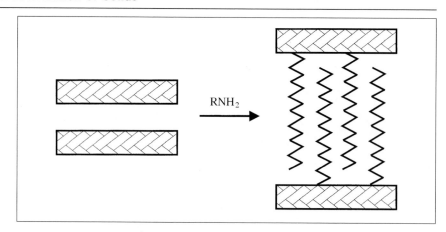

Figure 7.22 Concertina shape of RNH$_2$ inserted in TaS$_2$

Type 2: The Staging of Graphite

The structure of graphite is shown in Figure 7.23. The hexagonal rings of graphite form infinite layers separated by 3.5 Å. Adjacent layers are displaced relative to each other, so the repeat distance is *twice* the layer separation. The weak van de Waals interactions between the layers means that ions and molecules can be readily intercalated between them.

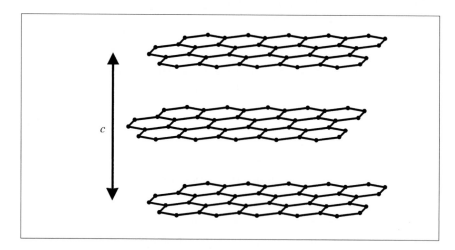

Figure 7.23 Graphite structure

By reaction with potassium vapour at 64 °C the compound C$_8$K is formed. The driving force for the reaction is the **electron transfer** between potassium and the graphite host.

While most compounds fill the sites between layers *randomly*, graphite does so in an *ordered* fashion. This process is known as **staging**. The first stage compound is C$_8$K, as shown in Figure 7.24, where potassium lies between *all* the carbon layers. The other compounds of potassium and carbon that exist are $C_{24}K$, $C_{36}K$ and $C_{48}K$. In $C_{24}K$, instead of removing potassium randomly from all the layers, alternate layers are emptied, and the remaining layers become two-thirds filled. Within the layers containing potassium ions, the arrangement of potassium ions is random. $C_{24}K$ is the second stage intercalate. $C_{36}K$ and $C_{48}K$ are the third and fourth stage intercalates, respectively, having potassium ions every third and fourth layer. The staging process is shown in Figure 7.25.

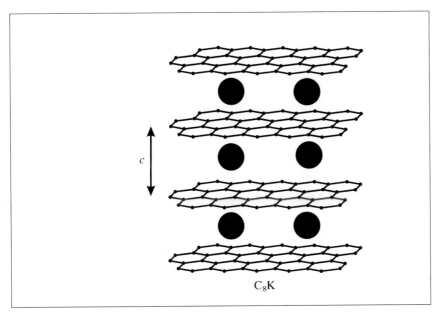

Figure 7.24 First stage intercalate

C_8K

Figure 7.25 The 2nd, 3rd and 4th stages of graphite

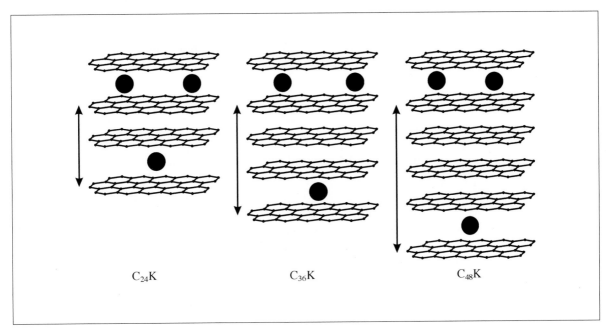

$C_{24}K$ $\qquad\qquad$ $C_{36}K$ $\qquad\qquad$ $C_{48}K$

Worked Problem

Q Show how the stoichiometry of the third stage intercalate can be calculated, using the information that every third layer is two-thirds filled.

A C_8K is the formula of a compound with filled layers. Therefore a compound with layers two-thirds filled would have a formula $C_8K_{2/3}$. The other two layers would not have any potassium in them. The repeat would therefore be: $C_8K_{2/3}C_8C_8$ which is $C_{24}K_{2/3}$. Dividing the formula through by two-thirds produces the stoichiometry $C_{36}K$.

Similar behaviour occurs with bromine, forming C_8Br, where the layers are positively charged. The intercalated materials appear gold coloured and they are metallic conductors.

7.4.2 Oxide Intercalates

A number of metal oxides undergo intercalation reactions, typically with *small* ions. They have the following features:

1. High oxidation state oxides with relatively open structures, *e.g.* V_2O_5, WO_3.
2. Only small ions such as Li^+ or H^+ can be inserted.
3. The metal cations in the host can normally be easily reduced, but reaction temperatures should be low to avoid complete reduction to the metal.

Lithium Reactions

Lithium can be inserted into a host structure by simple reactions. For example, grinding up LiI with V_2O_5 in a pestle and mortar at room temperature gives $Li_xV_2O_5$ (Figure 7.26). This produces a dramatic colour change as the intercalate forms and iodine is produced.

Lithium can also be intercalated using an electrochemical reaction in an organic solvent and also by using butyllithium.

Hydrogen Reactions

Hydrogen is another element with a small ion which can intercalate into

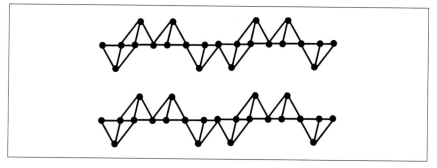

Figure 7.26 Layered structure of V_2O_5

oxide structures relatively easily. Several routes can be used to introduce the hydrogen into the structure:

1. Generation of hydrogen with metal/acid:

$$WO_3 + Zn / HCl \rightarrow xH + WO_3 + H_xWO_3$$

Hydrogen is created *in situ* to generate the intercalate. x can take values between 0 and 0.6. The parent material is a white insulator, and the colour of the intercalate ranges from blue to gold as x increases.

2. Activation with platinum; under a direct stream of hydrogen, using platinum as a catalyst.
3. Electrochemically, using sulfuric acid and a platinum electrode.

The last two methods are not so effective. For the tungsten compound above, the maximum value of x achievable is 0.3 by these methods.

7.5   Buckminsterfullerene, C_{60}

Buckminsterfullerene has been hailed as one of the most important discoveries of recent times. An accolade to its importance was the award of the Nobel Prize to its discoverers in 1996. It is peculiar in that it crosses all the borders of chemistry, from solid state, where it acts as an intercalation host, to organic/organometallic chemistry, where it acts as a ligand. Its unusual magnetic and electronic properties have also led to many physical studies. Interestingly, it is not named after its discoverers, but according to its shape. Buckminster Fuller designed a number of geodesic dome structures based on hexagons and pentagons, which are also apparent in C_{60} and the ordinary spherical football.

7.5.1 Preparation

Before 1985, six forms of carbon were known. Since the discovery of the closed-shell configuration of C_{60}, many other closed-shell forms have been found. C_{60} was initially prepared by vaporization and condensation of carbon in an inert atmosphere. However, it is even thought to be present in candle soot. The first crystals of C_{60} were grown from benzene solutions by the physicists Kratschmer and Huffman.[13] The experiment involved causing an arc between two graphite rods surrounded by helium, removing the carbon condensate and extracting it using benzene. Isolating pure C_{60} from the molecules of organic solvent contained in the crystals is extremely difficult, requiring repetitive sublimation.

The C_{60} molecule in the gas phase has perfect icosahedral, I_h, symmetry. The carbon atoms lie on a sphere of radius 3.55 Å, and each is in an identical environment with three neighbours. The structure is shown in Figure 7.27, and consists of 20 six-membered hexagonal rings and 12 five-membered pentagonal rings on the surface of a sphere. The pentagons are isolated from each other on the surface of the molecule to avoid the strain of two adjacent pentagons. There are two sorts of carbon–carbon bond, which are given the nomenclature 6:5 and 6:6. These numbers describe the bonds which connect the hexagons to the pentagons and hexagons, respectively. The 6:5 bonds are 1.45 Å in length, with the 6:6 slightly shorter at 1.40 Å. A ^{13}C NMR spectrum of C_{60} dissolved in methylbenzene shows each carbon is in an identical environment.

One of the most unusual properties of C_{60} is its solid state structure: how the molecules pack in a solid state lattice. Surprisingly, the mole-

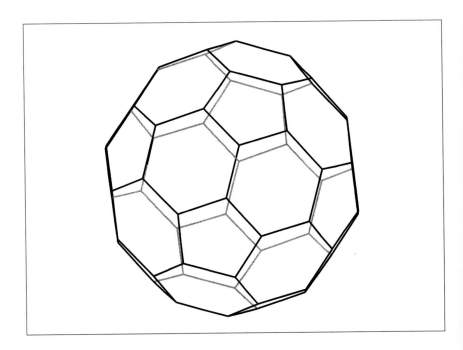

Figure 7.27 C_{60}

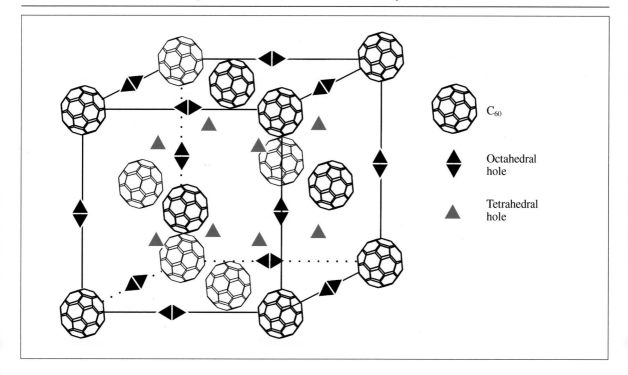

C_{60}

◆ Octahedral hole

▲ Tetrahedral hole

cules simply form a normal face-centred cubic lattice just as is observed for cubic close-packing of spheres (Figure 7.28).

The solid state NMR spectrum is even more perplexing, as Tycko *et al.* discovered.[14] Instead of a broad line, expected for a normal solid, C_{60} gives a sharp single line at 143 ppm without any magic angle spinning (Chapter 3). This means the solid is acting like a *liquid,* and the molecules must be tumbling on their lattice sites.

X-ray diffraction shows this to be true. The molecules have enough energy at room temperature to freely tumble on their lattice sites, behaving essentially like a molecule in solution.

Cooling C_{60} down to very low temperatures causes the molecules to lock in a particular conformation. Below 90 K, the static solid state pattern broadens into the spectrum expected for a normal solid with fixed lattice sites.

Figure 7.28 Face-centred cubic array of C_{60} creating octahedral and tetrahedral holes

7.5.2 Intercalation Chemistry

Solid C_{60} is a redox-active array, with relatively weak intermolecular forces (comparable to interlayer forces in graphite). Therefore, it is a potential host for intercalation chemistry, like graphite or the transition metal sulfides.

As with a normal face-centred cubic lattice, the C_{60} array generates tetrahedral and octahedral holes which can accept intercalated species

(Figure 7.28). If we consider C_{60} as spheres, this leads to the normal four spheres, eight tetrahedral holes and four octahedral holes per unit cell of a face-centred cubic lattice. The two holes have significant size differences, and as a result C_{60} has extensive intercalation chemistry. For example, caesium is too large to be accommodated in the tetrahedral holes but fits comfortably into the octahedral holes. Once C_{60} has been reduced it is known as a fulleride. The intercalated ions are mostly either metals of Group 1 or 2, where the former is most common.

Preparation of Fullerides

Fullerides can be prepared in a number of ways:

1. Directly using metal vapour and C_{60} in a sealed tube (as metal vapours are extremely volatile). This method is always employed when preparing compounds from Group 2.
2. Dissolution of the Group 1 metals in liquid ammonia and adding C_{60}. This route caused considerable confusion for a while, as ammonium ions as well as the Group 1 cations were being intercalated.

The size and number of intercalated cations have a pronounced effect on the electronic structure of C_{60}. C_{60} has a low-lying acceptor orbital which it uses for bonding. The band filling (by virtue of the anion) and the band width (by virtue of the separation of the molecules) are strongly affected by the ions intercalated. The maximum number of electrons that can be accepted by this acceptor orbital is 6; therefore, for the general formula $A_x C_{60}$, where A is an alkali metal, if A is less than 6 the intercalate should be conducting (partially filled bands).

Up to and including $x = 3$ the intercalates maintain the face-centred cubic structure and fill the octahedral and tetrahedral sites accordingly. However, the materials $A_6 C_{60}$ (where A = K, Rb, Cs) transform and form a new structure, consisting of a body-centred cubic array of C_{60} molecules with a cluster of metal atoms at the centre.

$A_3 C_{60}$: Conductors and Superconductors

This composition has produced the class of superconductors which has the highest T_c of any non-cuprate based system. The phases $K_3 C_{60}$, $Rb_3 C_{60}$ and $CsRb_2 C_{60}$ (maximum $T_c = 33$ K) have the intercalated face-centred cubic structure at all temperatures, with both octahedral and tetrahedral holes filled.

Although the basic structure is the same, the intercalated ions cause the molecule orientation to become fixed and non-rotating, with a six-membered ring pointing towards each cation. The effect of the intermolecular separation of the C_{60} on T_c is striking and is illustrated in Figure 7.29. This trend indicates that T_c is controlled by the density of states at the Fermi level. The higher density of states, the higher the T_c. The width of the conduction band and hence the density of states at the Fermi level depends on the orbital overlap. If the C_{60} molecules are close together, the overlap is good and the band is broad, giving a low density of states at the Fermi level. As the atoms are forced apart by larger intercalated species, the band narrows and the density of states at the Fermi level rises (Figure 7.30).

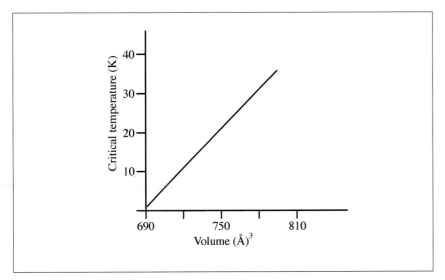

Figure 7.29 Change in critical temperature with volume of C_{60} unit cell

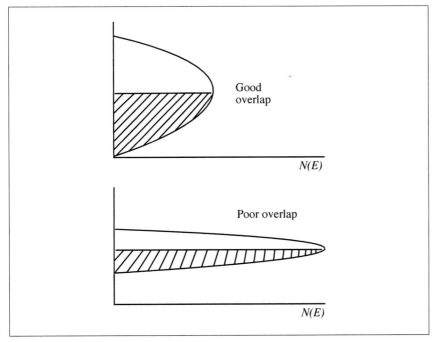

Figure 7.30 Schematic diagram of the density of states diagrams for C_{60}

Higher Fullerenes

Separation of the higher fullerenes is extremely difficult owing to the existence of multiple isomers. C_{70} is reasonably well known, and is similar to C_{60} with an extra band of 10 carbons added in the middle, making a rugby ball shape. Other materials such as C_{82} are known. The symmetry has been reduced to D_{5h} in these rugby ball-type materials, and intercalation experiments have shown the existence of K_4C_{70}. No superconducting higher fulleride has yet been discovered.

Further Reading

C. N. R. Rao, *The Chemistry of the High Temperature Superconducting Copper Oxides,* Springer, Heidelberg, 1991.

R. M. Barrer, *Zeolites and the Clay Minerals,* Academic Press, New York, 1978.

R. M. Barrer, *Hydrothermal Chemistry of Zeolites,* Academic Press, New York, 1982.

G. Englehardt and D. Michel, *High Resolution Solid State NMR of Silicates and Zeolites,* Wiley, New York, 1987.

D. W. Breck, *Zeolite Molecular Sieves: Structure, Chemistry and Use,* Wiley, New York, 1974.

H. H. Mal and K. H. J Buschow, *Intercalation Chemistry,* Academic Press, New York, 1982.

R. F. Curl and R. E. Smalley, *Science,* 1988, **242**, 1017.

H. Kroto, *Science,* 1988, **242**, 1139.

I. Hargittai, *Discoverers of Buckminsterfullerene,* in *The Chemical Intelligencer,* Springer, New York, 1995.

Problems

1. (a) Describe briefly how you would synthesize the solid state compound Na_3C_{60}.

(b) The ^{13}C MAS NMR spectrum of C_{60} recorded at room temperature shows a single narrow resonance, but as the temperature falls the spectrum broadens to a feature covering several kHz. Describe the origin of the differences in the spectra.

(c) Reaction of C_{60} with rubidium and caesium produces an <u>intercalation</u> compound with the formula, Rb_2CsC_{60}. Describe, with the aid of a diagram, why you would expect this compound to be a <u>type II superconductor</u>.

(d) Explain what is meant by the underlined terms in part (c).

2. Reaction of a mixture of yttrium oxide, barium carbonate and copper oxide in a stream of oxygen produces A. Elemental analysis of the oxide A gave 11.9% Y, 36.8% barium and 34.1% copper.
(a) Calculate the empirical formula of A, assuming the rest of the molecular % is oxygen.
(b) Determine the average oxidation state of the copper.
(c) If you brought a pellet of A at very low temperature towards a permanent magnet, what would you expect to happen and why?
(d) Draw the graph expected for the change in resistance with temperature for A. Explain the salient features of the graph, with the aid of diagrams where necessary. What type of superconductor would you expect A to be?

3. $Cu(OH)_2$ reacts with $Au(OH)_3$ in a 3:1 molar ratio to form A. The IR spectrum of A shows a strong absorption at 3400 cm^{-1}. 0.541 g (1 mmol) of A decomposes in a stream of hydrogen at 500 °C to produce B and water (0.162 g). Elemental analysis of the intermetallic B revealed 50.8% Au.
(a) Using the information given above, identify A and B.
(b) Given that B is a superconductor which conducts at liquid helium temperatures, predict a likely structure for B.

4. (a) Describe the structural features of graphite which make it suitable for intercalation reactions.
(b) How does the structure of the first stage intercalate of graphite differ from that of the parent material?
(c) Reaction of excess potassium with graphite produced the gold-coloured compound A which contained 28.9% potassium. Describe the type of reaction occurring, and explain the differences in appearance and conductivity of the new material compared with graphite.

5. (a) Briefly describe the important features of zeolites which make them suitable for ion exchange. Why do siliceous zeolites make poor water softeners?
(b) Describe two ways in which an H-zeolite could be prepared.
(c) Describe the two sorts of acid site which form in zeolites.
(d) Explain why zeolites can be used for shape-selective catalysis.

References

1. H. K. Onnes, *Akad. Van Wetenschappen,* 1911, **14**, 113.
2. J. G. Bednorz and K. A. Muller, *Z. Phys. B, Condens. Matter.,* 1986, **64**, 189.
3. M. K. Wu, J. R. Ashburn, C. J. Torrig, P. H. Hor, R. L. Meng, Z. J. Huang, Y. Q. Wang and C. W. Chu, *Phys. Rev. Lett.*, 1987, **58**, 908.
4. J. Bardeen, L. Cooper and J. R. Schrieffer, *Phys. Rev.*, 1957, **108**, 1175.
5. W. Meissner and R. Ochsenfield. *Naturwissenschaften,* 1933, **21**, 787.
6. C. P. Poole, H. A. Farach and R. J. Creswick, *Superconductivity,* Academic Press, San Diego, 1996, p. 22.
7. Y. Shimakawa, Y. Kubo and T. Manako, *Nature,* 1996, **379**, 53.
8. A. P. Ramirez, R. J. Cava and J. Krajewski, *Nature,* 1997, **386**, 156.
9. C. Zener, *Phys. Rev.*, 1951, **82**, 403.
10. A. F. Cronstedt, *Kongl. Svenka Vetenskaps. Acad. Hundlinger.,* 1756, **17**, 120.
11. C. Williams (ed.), *R. M. Barrer 1910–1996, Founding Father of Zeolite Science,* British Zeolite Association, University of Wolverhampton, 1999.
12. W. Löwenstein, *Am. Miner.*, 1954, **39**, 92.
13. W. Kratschmer, L. D. Lamb, K. Foshropoulos and D. F. Huffman, *Nature,* 1990, **629**, 354.
14. R Tycko, *Solid State Nucl. Magn. Reson.,* 1994, **3**, 303.

Answers to Problems

1. Figure 1.23

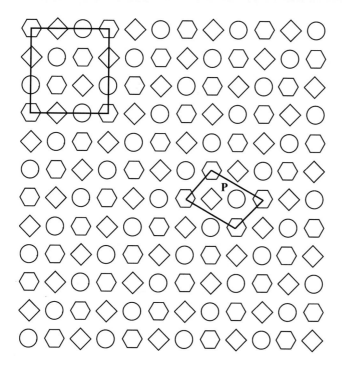

Figure 1.23 Primitive (P) and non-primitive unit cell

2. (i) Formula A = 4 atoms, C = 8/4 = 2 tetrahedral holes. Formula = A_4C_2 or A_2C. Coordination number of C must be four as it is in a tetrahedral hole. If all the holes were filled, then A would be eight coordinate; however, since only one quarter of the holes are filled, A must have a coordination number of two.

(ii) Using the same reasoning as above, the formula is A_4C and the coordination numbers are one and a half and six, respectively.

3. Figure 1.24. For a tetragonal unit cell $a = b \neq c$ and $\alpha = \beta = \gamma = 90°$.

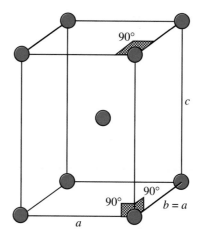

Figure 1.24 Body-centred tetragonal unit cell

4. Figure 1.25.

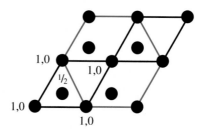

Figure 1.25 Outline of hexagon in hexagonal close-packing

5. There are two octahedral holes within the unit cell, each counting one. Eight tetrahedral holes lie on the edges, each contributing one quarter, and two lie within the unit cell, each contributing two (four in total). There are eight atoms on the corners, each contributing one eighth, and one within the unit cell, contributing one. This gives a ratio of octahedral holes:tetrahedral holes:atoms of 1:2:1.

6. Figure 1.26. The coordination number of the tungsten atoms is only eight, so it cannot be close packed as the coordination number must be 12. The ionic radius is just half the distance between the two closest atoms. This is just one quarter of the body diagonal of the cube = $\frac{1}{4}\sqrt{3}a$.

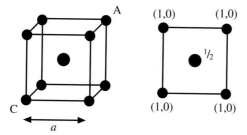

Figure 1.26 Body-centred cubic unit cell of β-tungsten

7.

$$19.5 = \frac{\dfrac{R.A.M}{N_A} \times 2}{(3.15 \times 10^{-8})^3}$$

Therefore R.A.M. = 183.5.

8. (i) Hexagonal close packed: two-layer repeating sequence, ABABAB . . .
Cubic close packed: three-layer repeating sequence, ABCABCABC- . . .
Both have a coordination number of 12, with 12 sphere nearest neighbours.
(ii) Figure 1.27.

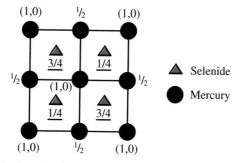

▲ Selenide
● Mercury

Figure 1.27 Projection of HgSe forming the zinc blende structure

(iii) Fluorite is cubic close-packed structure with all tetrahedral holes filled, whereas no hexagonal close-packed structure exits with all tetrahedral holes filled.

9. (i) A tetrahedral hole is created by four spheres placed symmetrically, three in the plane and one in the depression produced by them. A hole is created between them since the round spheres do not fill all the space. The hole can be filled by other ions to form compounds.
(ii) Lithium is much smaller than the oxide ion so the positions of cations and anions are reversed compared to the fluorite structure, *i.e.* where the smaller ions fill the holes and the larger ions make up the lattice.
(iii) Coordination number of Li = 4 and O = 8:

$$Li = 8 \times 1 = 8; \ O = (8 \times 1/8) + (6 \times {}^1\!/_2) = 4$$
Formula = Li_2O and there is Li_8O_4 [4(Li_2O)] in the unit cell, so $Z = 4$.

10. $r^+/r^- = 0.611$ for FeO. There the material is likely to contain six-coordinate ions. This means the halite structure would be suitable.

Chapter 2

1. (i) Using the Kapustinskii equation, $V = 2$, Ca and O both have a charge of 2. Since the cell edge of CaO is 483 pm and it forms the halite structure, then the shortest distance between cation and anion is 241.5 pm. Then

$$U_L = \frac{(1.214 \times 10^5) \times V \times z_1 \times z_2}{r_1 + r_2} \left[1 - \frac{34.5}{r_1 + r_2} \right]$$

$$= \frac{(1.214 \times 10^5) \times 2 \times 2 \times 2}{241.5} \left[1 - \frac{34.5}{241.5} \right]$$

$$= 3447 \text{ kJ mol}^{-1}$$

(ii) See Figure 2.9.

$$\Delta H_f = 590 + 1100 + 249 - 142 + 844 - 3447 = -806 \text{ kJ mol}^{-1}$$

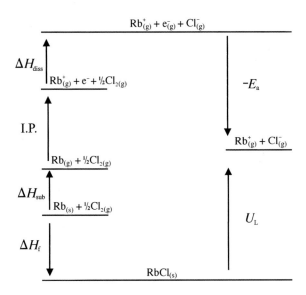

Figure 2.9 Born–Haber cycle for the formation of RbCl

(iii) Ionization potentials and electron affinities are defined at 0 K and thermochemical cycles are determined at 298 K.

(iv) The second ionization potential for calcium is much greater that the first despite forming a closed shell configuration because it involves removing an electron from a positively charged ion. The electrostatic interaction is therefore much greater and the electron is held closer to the nucleus.

2. From data tables, the ionic radius for Sr^{2+} = 113 pm and for O^{2-} = 140 pm. This gives an approximate minimum cation–anion distance of 253 pm. Using the Kapustinskii equation as in question 1 gives U_L = 3315 kJ mol^{-1}.

3. Using the Kapustinskii equation as in question 1 gives 530 kJ mol^{-1}.

4. The exact value depends on the data book used, but using a similar Born–Haber cycle as given in Figure 2.6 and substituting Rb for Cl gives a value in good agreement with the ionic model (± 10 kJ mol^{-1}).

5. $3U_L(AlX) + I_2 + I_3 - U_L(AlX_3) - 2I_1(Al) - 2\Delta H_{at} = \Delta H_{dec}$.
Using the data given for fluoride = 3(910) + 1816 + 2743 – 6380 –

$2(577) - 2(324) = -893$ kJ mol^{-1}. Using the data given for iodide = $3(696) + 1816 + 2743 - 4706 - 2(577) - 2(324) = +159$ kJ mol^{-1}. The most stable species in each case is AlF$_3$ and AlI as the ΔH_{dec} is exothermic for the fluoride but endothermic for the iodide.

6. (i) The double bond is strong for C=O as the elements are in the same period and the pπ–pπ interaction is strong. For Si=O this interaction is weaker as the elements are in different periods. However, the low-lying d-orbital on silicon strengthens the single bond of Si–O by pπ–dπ interaction. C has no low-lying d-orbital, and the bond is weak.

(ii) Heterolytic bond enthalpies are generally larger than homolytic bond enthalpies because the bonds are strengthened by the electrostatic interaction. Since C and F have different electronegativities, the bond is strong.

(iii) The N–N bond is weaker than the F–F single bond as the lone pairs on the nitrogen destabilize the bonding interaction by electron repulsion.

(iv) The silicon is larger than carbon and has a low-lying empty d-orbital. This means it can be attacked by nucleophiles such as water. The Si–Cl bond is weaker than the Si–O bond. In contrast, the C is very small and has no empty orbitals and is not easily attacked. The C–F bond is stronger than C–O, and the compound is not easily hydrolysed.

Chapter 3

1. See Figure 3.24:

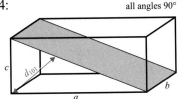

2.

$$\sin^2\theta = \frac{\lambda^2}{4a^2}(h^2 + k^2 + l^2)$$

$\sin^2\theta$ for the (301), (400) and (111) = 0.0386, 0.0617 and 0.0154, respectively; therefore $2\theta = 22.661$, 28.765 and 14.257, respectively.

3. Since the unit cell is body-centred cubic, $h+k+l = 2n$. The first ten reflections are therefore
110, 200, 211, 220, 310, 222, 321, 400, 411 and 420.

4.

Angle	θ	$\sin^2\theta$	ratio	×3	h,k,l
22.593	11.297	0.0384	1	3	111
26.145	13.073	0.0512	1.33	4	200
37.311	18.656	0.1023	2.66	8	220
44.059	22.030	0.1406	3.66	11	311
46.129	23.065	0.1535	4	12	222
53.792	26.896	0.2046	5.33	16	400

The lattice type is face-centred cubic (thirds in the initial ratio). Using the last reflection to calculate the lattice parameter gives 8.401 Å.

5. (a) The potassium cation and the chloride anion are isoelectronic and will be indistinguishable in X-ray diffraction, since the X-rays are scattered by the electron cloud. In a similar way to MgO (given in the text), the unit cell observed by X-rays is a primitive cubic with a lattice parameter half that expected for KCl. No restrictions on the observed reflections.
(b) The potassium cation and the chloride anion have very different scattering lengths in neutron diffraction and the true face-centred cube will be observed. Hence h,k,l must be all odd or all even.

6.

$$\frac{24.2}{\text{mass SrFeO}_{3-x}} = \frac{23.34}{\text{mass}\left(\frac{1}{2}\text{Fe}_2\text{O}_3 + \text{SrO}\right)}$$

Therefore

$$\frac{24.2}{143.467 + (3 - x(16))} = \frac{23.34}{183.467}$$

On rearranging

$$3 - x = \frac{24.2(183.467) - 23.34(143.467)}{23.34(16)}$$

Therefore $3 - x = 2.92$, which gives $x = 0.08$.

7. The 110 would be observed for both body-centred cubic ($h+k+l = 2n$) and primitive (no restrictions) systems, but not for a face-centred system (as h,k,l must be all odd or all even). However, since the 234 is not observed, this implies the system must be bcc.

8. X-rays are scattered by electrons. Since this compound contains one heavy element (Sr) and two light ones (Li, H), in an X-ray diffraction pattern most of the scattering will be generated by the Sr:

2θ	θ	$\sin^2\theta$	ratio	h,k,l
23.25	11.63	0.040	1	100
40.84	20.42	0.121	3	111
47.53	23.77	0.162	4	200
53.56	26.78	0.203	5	210
69.49	34.75	0.324	8	220
74.38	37.19	0.365	9	300

The lattice is primitive as there are no systematic absences. Using the last reflection gives 3.822 Å.

Chapter 4

1. Yttrium barium copper oxide has been prepared in a number of ways using both conventional ceramic and co-precipitation methods. The classic laboratory method uses yttrium oxide, barium carbonate (not barium oxide as it is hygroscopic) and copper oxide.

2. The reaction temperature could be lowered by using the precursor method of preparation, *e.g.* dissolve cobalt acetate and aluminium acetate in acetic acid. Evaporate to dryness and heat the residue.

3. Lithium titanate could be prepared by using a sealed tube method. As the alkali metals can react with glass, the tube must be coated with graphite by decomposition of an organic solvent on the surface. Then a mixture of lithium oxide and titanium dioxide could be weighed out in a glove box, placed in the tube and the tube evacuated. Once sealed off, the whole tube could be heated in a furnace. Alternatively, a mixture of lithium carbonate (in excess) and titanium dioxide could be pressed into pellets and heated. The required excess of lithium carbonate would have to be determined by trial and error.

4. This reaction just requires the reactants to be mixed in the correct molar proportions to give the right ratio and to make sure there is no oxygen left over. The reactants would then be reacted in a sealed tube, in a similar way to that described in question 3, to prevent aerial oxidation:

$$3V_2O_{5(s)} + 4V_{(s)} \rightarrow 5V_2O_{3(s)}$$

5. This reaction uses cobalt oxide and alumina as starting materials. Choice of an alumina crucible would therefore by unsuitable. This is because raising the temperature sufficiently to get the alumina starting material to react would also allow the crucible to react. Any other reaction vessel with a melting point higher than 1200 °C would be suitable (Table 4.1).

6. Solid B and solid AB using the phase diagram.

7. A single crystal of zeolite could be prepared hydrothermally using slow cooling from the reaction temperature.

Chapter 5

1. This should be the same diagram as silicon doped with aluminium (Figure 5.9iii), since germanium is in Group 14 and gallium in Group 13. The type of charge carrier is a positive hole and the semiconductor is p-type.

2. In the carbon layers in graphite, each carbon is sp^2 hybridized. This arrangement uses *three* of the *four* available orbitals on carbon for bonding. Each carbon centre is trigonal planar, which produces the layered structure. However, this also means there is one p-orbital (containing one electron) remaining on each carbon which is *not* involved in bonding. These orbitals can overlap with p-orbitals on the carbons in other layers to form bands. In diamond the carbons are sp^3 hybridized. All the orbitals which contain the outer electrons are used for bonding and there are no orbitals remaining which contain outer electrons to form a band. Therefore diamond is insulating.

3. Increasing the number of anions creates cation vacancies (*i.e.* the formula ZnO becomes $Zn_{1-x}O$ as there are less cations than anions). There are positive holes created where the cations should be.

4. (a) No, as an insulator is just a semiconductor with a very large band gap.

(b) Yes, the semiconductor has a small band gap compared with an insulator and will conduct at a lower temperature.

(c) The band gap will be smaller, but the actual picture will be very similar.

5. (a) MnO will antiferromagnetically order using the superexchange mechanism (Figure 5.21).

(b) Co metal has unpaired d-electrons and orders ferromagnetically (Figure 5.20).

6. (a) Initially, conductivity is metal-like. Once beyond the critical temperature the conductivity increases dramatically.

(b) Resistance falls with falling temperature, just like a metal, until the critical temperature is passed. Below this point, the resistance falls away faster, but gradually (as it is a type 2 superconductor and in the vortex state). Finally, the resistance reaches zero and all of the material is superconducting.

Chapter 6

1. Potassium chloride will have Schottky defects. Exposing the sample to excess potassium vapour causes potassium to be inserted on to the empty sites. The colour results from electron trapping (*i.e.* K is inserted as a cation and the electron is trapped in the lattice) in an empty site. This is an extrinsic defect.

2. (a) Number of moles of Ce = 0.025. Number of moles of deuterium (D_2) = 600/24000 = 0.025. Therefore A = CeD_2, which is likely to form the fluorite structure by analogy with UO_2.

(b + c) B forms by insertion of anions on to an interstitial site in an extrinsic defect. It forms by the clustering of vacancies/interstitials, such that two interstitials are incorporated for every vacancy (by analogy with UO_{2+x}). An extra 75 cm^3 of D_2 corresponds to 0.003125 moles of D_2, which gives the formula $CeD_{2.25}$.

3. Defects are incorporated into the halite structure by formation of vacancies on the normal lattice sites. When $x = 0.7$, there are simply vacancies in the oxide sublattice. When $x = 1.25$, which can be rewritten by dividing the formula by 1.25 as $Ti_{0.8}O$, the vacancies are on the cation sublattice.

4. An intrinsic defect does not affect the stoichiometry, *e.g.* the

Schottky defects in NaCl where the ion vacancies occur in pairs. Extrinsic defects do affect the stoichiometry and are often formed by exterior influences on the lattice, *e.g.* by heating NaCl in potassium vapour. F centres are formed as the potassium is incorporated on to empty cation sites.

5. (a) This is similar to the formation of iron carbide. Hydrogen is incorporated into the structure by filling the empty octahedral sites created by the face-centred cubic array of palladium atoms. (b) This is an interstitial solid solution.

6. Formula $= 9 \times MoO_3 + 1 \times MoO_2 = Mo_{10}O_{29}$.

7. (a) Intrinsic defects: Schottky.
(b) Extrinsic defects: oxygen vacancies.
(c) Intrinsic defects: Frenkel.

8. The difference in size between the divalent calcium cation and the trivalent yttrium cation is less than 10%. This means that a substitutional solid solution *should* form. The fluorite structure would probably accommodate the difference in charge between the two cations by incorporation of interstitial anions.

Chapter 7

1. (a) Combination of the pure fullerene (repetitively sublimed) with sodium, in the correct molar proportions in a graphite coated silica tube. Evacuate tube and seal and heat to 70 °C.
(b) At room temperature the molecules of fullerene are tumbling on their lattice sites. The spectrum produced is therefore similar to that achieved for a liquid (where molecules tumble in solution). As the temperature falls the molecules lose energy and lock on their lattice sites. The spectrum broadens and becomes the same as a normal solid (Chapter 3: MAS NMR).
(c) Similar to type 1, except that the transition to the superconducting state is a gradual slope rather than a sharp fall. The large molecules force the C_{60} molecules apart and lead to poor overlap of the orbitals, giving a high density of states at the Fermi level (Figure 7.30).
(d) A superconductor is a compound which conducts with no resistance below a certain temperature (critical temperature). The conducting current moves with the lattice vibrations in a concerted

process such that there is no resistance. Type 2 superconducters allow partial penetration of the magnetic flux at intermediate values of the applied field, which leads to a gradual loss of resistance. This means there are both superconducting and non-superconducting areas in the same sample (vortex state). An intercalation compound forms without major rearrangement of the host lattice, which has strong (intra) and weak (inter) interactions and vacant sites which can accept ions/atoms.

2. (a) By division of the percentages by the relative molecular weights of yttrium, barium and copper, a ratio of 1:2:4 is obtained. By assuming the rest of the mass is oxygen, a formula of $YBa_2Cu_4O_8$ is achieved, *i.e.* Y (11.9/88.9 = 0.134), Ba (36.8/137.3 = 0.268), Cu (34.1/63.5 = 0.534) and O (17.2/16 = 1.075) gives 0.134:0.268:0.534:1.075 = 1:2:4:8.

(b) Average copper oxidation in state n in Cu^{n+}: $(8 \times -2) + 3 + (2 \times 2) + (4 \times n) = 0$. Therefore $n = 9/4 = 2.25$.

(c) Copper oxidation state is non-integral; therefore a possible superconductor. A permanent magnet would be repelled from the superconductor at low temperature. A surface current is set up on the surface of the superconductor to exactly oppose the external magnetic field and exclude magnetic flux from the sample (Meissner effect). Beyond a critical field the effect can no longer be overcome and the material would return to the normal state.

(d) See Figure 7.31. See question 1d for explanation of type 2 superconductivity.

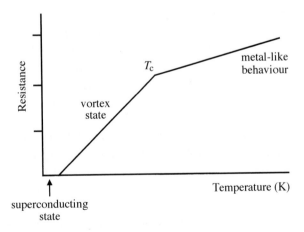

Figure 7.31

3. (a) IR absorption: characteristic of OH. On reduction in hydrogen, A produces 0.162 g of water which is exactly 9 mmol. This implies that A contains 9 moles of O, probably as OH. From the reaction stoichiometry, 3 moles of Cu and 1 mole of Au are probably in the product. Since 1 mmol of A weighs 0.541 g, the RMM of A should be 541. Now $3 \times$ RMM (Cu) + $1 \times$ RMM (Au) + $9 \times$ RMM (OH) = RMM A. Therefore RMM A = $(3 \times 63.54) + 196.97 + (9 \times 17) = 540.6$. Thus A = $AuCu_3(OH)_9$. B is formed from the reduction of A in hydrogen. B is an intermetallic. This implies loss of OH to give water with the formation of $AuCu_3$. If B is 50.8% Au then it must be 49.2% Cu. Therefore Au $(50.8/196.97 = 0.257)$ and Cu $(49/2/63.54 = 0.774)$; $0.257:0.774 = 1:3$, *i.e.* $AuCu_3$.

(b) Since B is superconducting and has the formula A_3B, it implies the structure given in Figure 7.4, *i.e.* a body-centred cube of Au atoms with two equidistant Cu atoms on the middle of each face, forming three non-intersecting chains.

4. (a) Graphite consists of strong covalently bonded layers of sp^2 hybridized C atoms with weak van der Waals forces between the layers. The layer structure provides an array of empty lattice sites and the layers are redox active.

(b) In the parent material the repeat distance is twice the distance between the layers, as the layers are askew. In the first stage intercalate the layers are aligned and the repeat distance is the same as the interlayer separation.

(c) 28.9% K and therefore 71.1% C. This gives the formula C_8K from: K $(28.9/39.1 = 0.739)$ and C $(71.1/12 = 5.925)$; thus $0.739:5.925 = 1:8$. This is a redox reaction where the potassium is oxidized and the graphite is reduced. The driving force for the reaction is the loss of the electron from the potassium. The intercalated material has a conductivity 1000 times that of graphite and shows true metallic conductivity, where the resistance rises with temperature.

5. (a) A zeolite consists of a three-dimensional framework which is constructed from vertex-linked silicate and aluminate tetrahedra and voids/channels. For every aluminium there will be one monovalent cation for change balance and hydrating water molecules. Siliceous zeolites make poor water softeners as they contain few cations, since cations are occluded for charge balance to compensate for the aluminium substitution, *i.e.* where aluminium is trivalent and silicon is tetravalent.

(b) Acidic zeolites can be prepared (1) by exchange in dilute acid, (2) by exchange with an ammonium salt and then heating to remove the ammonia.

(c) Brønsted and Lewis acid sites (Figure 7.19).

(d) The windows and channels in zeolites result in a restriction on size and shape. Molecules of the wrong geometry or size cannot pass through the zeolite. This means if molecules are generated *in situ* inside the zeolite of the wrong geometry, they can often be converted into other isomers using the acid sites.

Subject Index